IMAGE POLITICS

IMAGE POLITICS

The New Rhetoric of Environmental Activism

KEVIN MICHAEL DELUCA

Routledge
Taylor & Francis Group

NEW YORK AND LONDON

PHOTO CREDITS: The photographs appearing on pages 2, 33,53, and 99 are reproduced by the courtesy of Greenpeace International with special thanks to Elkie Jordans of the Photo and Video Library. The photographs appearing on pages 7, 55, 105, 117, and 160 are reproduced by the courtesy of Northcoast Earth First! with special thanks to Michael Avcollie. The photographs appearing on pages 9 and 106 are reproduced by the courtesy of Environmental Images and the photographer, Elizabeth Feryl. The photographs appearing on pages 11 and 162 are reproduced by the courtesy of the photographer, Craig R. Braack, with thanks. The photograph appearing on page 75 is reproduced by the courtesy of Kentuckians for the Commonwealth and with special thanks to Jerry Hardt. The photographs appearing on pages 110 and 116 are reproduced by the courtesy of ABC News with thanks to Nicole Halpern for her help. The photograph appearing on page 125 is reproduced by the courtesy of Kelpie Wilson of Siskiyou Project with thanks.

LAWRENCE ERLBAUM ASSOCIATES, INC.

First Published by
Lawrence Erlbaum Associates, Inc., Publishers
10 Industrial Avenue
Mahwah, New Jersey 07430

Transferred to Digital Printing 2009 by Routledge
270 Madison Ave, New York NY 10016
2 Park Square, Milton Park, Abingdon, Oxon, OX14 4RN

Cover layout by Kathryn Houghtaling Lacey

Library of Congress Cataloging-in-Publication Data

DeLuca, Kevin Michael.
Image Politics: the new rhetoric of environmental activism / Kevin Michael
    DeLuca.
        p. cm.
Includes bibliographical references.
ISBN 1-57230-461-8 (hc.)
ISBN 0-8058-5848-2 (p)
I. Environmentalism. 2. Mass media and the environment. 3. Rhetoric   I. Title.
GE195.D45   1999
363.7'0525—dc21   99-19839
                                                                            CIP

Publisher's Note
The publisher has gone to great lengths to ensure the quality of this reprint
but points out that some imperfections in the original may be apparent.

To Grandpa for building the foundation I stand on
To Dad for being my greatest teacher
To Michael and Silas with all my love

Communication is the problem to be answered.
—10CC "The Things We Do for Love"

We need to find a way of thinking about opinion formation that recognizes the distinctiveness of a process that relies more on the image than the word, a process that is more figural than discursive, a process that creates "meanings" in which the cognitive content is underarticulated and is dominated by highly charged visual components.

—Andrew Szasz, *Ecopopulism*

ACKNOWLEDGMENTS

Writing this book has been a collaborative effort in the best sense. It is with pleasure that I acknowledge the help of many people. The references should make clear my prinicipal intellectual debts, but such debts exceed a simple list. When I was an undergraduate, William Johnstone introduced me to the life of the mind and encouraged me to pursue it. Ian Angus taught me the way of thinking. The intellectual community at the University of Iowa immersed me in rhetoric and deconstruction and enabled me to write the dissertation that later evolved into this book. I am especially grateful to John Peters for his guidance and care, Michael McGee for his inspiration and cultivation, Barbara Biesecker for her teaching and thinking, and Bruce Gronbeck for his questions. My former colleagues at Penn State were wonderfully supportive. Steve Browne has my special thanks for his extraordinary efforts on my behalf. Staffs at both the University of Iowa and at Penn State, especially Carol Schrage and Rita Munchinski, have made my life easier in innumerable ways. Gretta Armstrong and Denise Waggoner provided needed assistance with the computer technologies that have transformed the shape of our lives. Research assistants Davin Grindstaff and Christine Harold helped prepare the manuscript. The editors of the Revisioning Rhetoric series, Karlyn Kohrs Campbell and Celeste Michelle Condit, have been insightful, challenging, and supportive. Their engagement with my work has made this a much better book. Peter Wissoker and the staff at Guilford have been friendly experts at the many tasks of producing the book. They have made it a thoroughly enjoyable experience.

In addition to being an intellectual adventure, writing a book has been an emotional and spiritual journey that required the help of family and friends. My brother Scott and my aunt Ginny have always been there for me—I thank them for that invaluable gift. I wish to thank my mother June, for financial support and, more importantly, for being a part of my life. Friendship is a great gift and my friends have been generous with their love, care, and ideas. I owe Carol Corbin, Fernando Delgado, Pua Aiu, Jean Retzinger, Isabel Costa, Leah Ceccarelli, and Laura and Vico Guerrero more than they know. In addition to being great friends, John Delicath and Larry Rifkin have been dedicated readers and generous critics. Finally, to Ginny I owe great thanks for sharing the sublimity of life and love.

PREFACE

A *preface* would retrace and presage here a *general* theory and practice of deconstruction, that strategy without which the possibility of a critique could exist only in fragmentary, empiricist surges that amount in effect to a non-equivocal confirmation of metaphysics. The preface would announce in the future tense ("this is what you are going to read") the conceptual content or significance . . . of what will *already* have been *written*. And thus sufficiently *read* to be gathered up in its semantic tenor and proposed in advance. From the viewpoint of the foreword, which recreates an intention-to-say after the fact, the text exists as something written—a past—which, under the false appearance of a present, a hidden omnipotent author (in full mastery of his product) is presenting to the reader as his future. Here is what I wrote, then read, and what I am writing that you are going to read. After which you will again be able to take possession of this preface which in sum you have not yet begun to read, even though, once having read it, you will already have anticipated everything that follows and thus you might just as well dispense with reading the rest. . . . This is an essential and ludicrous operation.

—Jacques Derrida, *Dissemination*

The above quote from Derrida (1981a, p. 7) was my original idea for a preface—the entire preface. I was gently advised not to be quite so self-effacing. Given the slightly unorthodox form of the body of this text, it was sound advice. Still, Derrida's words on the preface highlight some of the themes running through this volume. These include a certain questioning of transmission, authorial power, intentionality, and textual authority, and a recognition of dissemination and the openness of meaning. For those who wish to avoid an authorial attempt at a partial closure, feel free to skip the rest of the preface.

This project is rooted in a conviction that industrial civilization is headed for disaster due to its perspectives on nature, humanity–nature relations, and how those relations should be mediated by technology. The need for radical change I take as a given. While practices and ideas are inextricably entwined, I start from the need to change worldviews or hegemonic discourses. Initially, I naively assumed that better ideas will out in the end, despite having read Thomas Kuhn's work. Very quickly, though, I realized the centrality of politics. Working from what is for me a native position, that there is no inherent meaning or telos for life or the universe, I approached politics and history as open, not as the unfolding of a scripted drama based on either essential foundations or teleological evolutionary laws. Laclau and Mouffe's work on articulation (see, e.g., 1985) enabled me to start to make sense of how social change is enacted and perpetuated. They also introduced me to a discursive theory of politics and social change. Something was still missing, though. If nothing else, Laclau and Mouffe's work seemed a bit distant from the daily practices of radical environmental groups. How are people persuaded, moved? In a word: rhetoric. Through rhetorical practices, people construct, perpetuate, and transform identities, discourses, communities, cultures, and worldviews.

My next problem was how to make sense of the primary rhetorical tactic of radical environmental groups: staging image events for mass media dissemination. Such a tactic falls outside the domain of a rhetoric traditionally conceived. Indeed, image events tend to slip the bounds of conventional conceptions of politics, social movement theory, and communication theory, as well as rhetoric. Yet images are clearly central to the practice of environmental politics today and, arguably, have been since Carlton Watkins' 1861 landscape photographs were used to help persuade the U.S. Congress and President Lincoln to designate Yosemite the world's first wilderness park. Instead of progressing in a linear fashion in a march to the Truth of image events through the deployment of ready-to-hand theoretical tools, I perform three meditations, designed both to offer insights about image events and to rethink conventional theoretical perspectives. After introducing image events, their performers, and their stage in Chapter 1, in Chapters 2 and 3 I meditate on image events from the perspective of the rhetorical theory of social movements. Conversely, I rethink the rhetoric of social movements in light of image events. Chapter 4 is a meditation on image events in the context of postmodern politics. In many ways, it is an affirmation of the political potential of postmodernism and image events.

An extended meditation in Chapters 5 and 6 deploys and questions the transmission model of communication, the putative heart of communication theory, rhetorical theory, media theory, and dominant ideology analyses of the culture industries. In Chapter 5 I question how much we should celebrate when image events must operate in an electronic public sphere dominated by a few large corporations. Adopting a dominant ideology frame, I perform close readings of image events embedded in television news broadcasts. In Chapter 6 I resuscitate hope with the help of cultural studies and deconstruction, rereading one of the television news broadcasts through the filters of audience research and dissemination. I end in Chapter 7 by considering the roles of critical rhetoric and the critical rhetorician in a postmodern world.

A note on tone. As many have noted, we live in the midst of monumental changes. We are burdened and blessed with the old Chinese curse, "May you live in interesting times." The Chinese character for crisis (*wei ji*) means both danger and opportunity. Enough has been written about the dangers of image events, the mass media, radical environmental groups, incivility, irrationality, micropolitics, the loss of belief in grand narratives, and image politics. Infused by an irrational feeling of hope, herein I explore the opportunities.

A final note on form. The form is designed to question, interrupt, and disrupt a certain drive to clarity, transparency, and a transmission of authorial intentions in a translucent text. I do not think the meditations offer a progression and, indeed, do not think they must be read in order. I also do not think I am giving answers. Instead, I hope this open form of meditations resembles the image events themselves, dense fragments floating in and out of the disparate discourses and contexts of a heteroglossic public sphere, open to many interpretations and having indeterminate effects. In the end, I hope to be raising questions and offering possibilities, much in the spirit of Elizabeth Grosz:

> In refusing to seek answers, and in continuing to pose questions as aporias, as paradoxes—that is, to insist that they have no readily available solutions—is to face the task, not of revolution, i.e., the overthrow of the old (whether capitalism, patriarchy, binary oppositions, or prevailing models of radicality) but, less romantically or glamourously, endless negotiation, the equation of one's life with struggle, a wearying ideal but one perhaps that can make us less invested in any one struggle and more capable to bearing up to continuous effort to go against the relentless forces of sameness, more inventive in the kinds of subversion we seek, and more joyous in the kinds of struggle we choose to be called into. (1995, p. 6)

CONTENTS

IMAGE POLITICS

CHAPTER 1

MAKING WAVES

On June 27, 1975, 50 miles off the coast of California, the Soviet whaling ship the *Vlastny*, armed with a 90-millimeter cannon loaded with a 160-pound exploding grenade harpoon, departs from the factory ship *Dalniy Vostok* in pursuit of sperm whales. Unlike any previous hunt, though, the *Vlastny* finds itself pursued by six Greenpeace activists in three Zodiacs (inflatable rubber dinghies) "armed" with one film camera and intent on confronting the whaler and intervening on behalf of the whales. One Zodiac, bobbing in and out of sight on the rough swells, manages to position itself between the harpoon ship and the nearest whale. The two activists in the Zodiac are betting that the whalers will not risk killing humans in order to kill whales. They lose. Without warning, the whalers fire over the heads of the activists, striking the whale. The steel harpoon cable slashes into the water less than 5 feet from the Zodiac.

Though Greenpeace's direct action failed in its most immediate goal of saving the whale, it succeeded as an image event.[1] Greenpeace caught the confrontation on film, and it became the image seen around the world, shown by CBS, ABC, and NBC News and on other news shows spanning the globe. For Robert Hunter, director of Greenpeace at the time and one of the activists in the path of the harpoon, Greenpeace had succeeded in launching a "mind bomb," an image event that explodes "in the public's consciousness to transform the way people view their world" (1971, p. 22). The consequence of this image event for Greenpeace was, as Hunter observed, that "with the single act of film-

Two Greenpeace activists in an inflatable Zodiac confront a Soviet whaling fleet. In capturing this initial protest on video and then disseminating it to news organizations, Greenpeace succeeded in igniting international indignation, jumpstarting the campaign to ban whaling, and establishing themselves as a force in the international public sphere.

ing ourselves in front of the harpoon, we had entered the mass consciousness of modern America" (1979, p. 231).

This opening act of Greenpeace's "Save the Whales" campaign echoed Greenpeace's founding act 4 years earlier, which also failed as a direct action but succeeded as an image event. Expatriate Americans and Canadians upset with the U.S. nuclear testing program chartered two boats to travel to Amchitka, one of the Aleutian Islands, in order to bear witness to and protest a scheduled underground nuclear explosion there. Underfunded and poorly equipped, the Greenpeacers were over 1,000 thousand miles away when the test took place. Though Greenpeace failed to stop that blast, the resulting publicity (two Canadian journalists were also Greenpeace crew members) generated a groundswell of protest and forced the U.S. Atomic Energy Commission to announce 4 months later that it was ending testing in the Aleutians for "political and other reasons" and returning Amchitka to its use as a bird and sea otter refuge (Brown and May, 1991, p. 15). Though at the time the Greenpeace crew members were unaware that "while a battle had been lost, the war had been won" (Hunter, 1979, p. 113), they have since learned that with image events it "is not whether they immediately stop the evil—they seldom do. Success comes in reducing a complex set of issues to symbols that break people's comfortable equilibrium, get them asking whether there are better ways to do things" (Veteran Greenpeace campaigner, quoted in Horton, 1991, p. 108).

Since 1971 Greenpeace has performed thousands of image events in support of issues ranging from whaling, to nuclear testing, to the siting of hazardous waste incinerators. Greenpeace activists have steered rubber rafts between whaling ships and whales, chained themselves to harpoons, spray-painted baby harp seals to render their pelts worthless, plugged waste discharge pipes, simultaneously hung banners from smokestacks in eight European countries in order to create a composite photograph that would spell out "STOP" twice, dressed as penguins to protest development of Antarctica, delivered a dead seal to 10 Downing Street (home of the British prime minister), and used drift nets to spell out "Ban Drift Nets Now" on the Mall in Washington, DC (Brown and May, 1991). The effects have been stunning. Greenpeace has parlayed the practice of creating image events as their primary form of rhetorical activity into the largest environmental organization in the world, reaching heights of almost five million members and gross revenues of $160 million (Horton, 1991, p. 44).

These tactical image events have driven numerous successful campaigns that have resulted in the banning of commercial whaling, harvesting of baby harp seals, and ocean dumping of nuclear wastes; the establishment of a moratorium in Antarctica on mineral and oil exploration and their extraction; the blocking of numerous garbage and hazardous waste incinerators; the requirement of turtle excluder devices on shrimp nets; the banning of the disposal of plastics at sea by the United States; and much more.

The vehemence of the counterresponse also testifies to the power of Greenpeace's image events. French commandos boarded a Greenpeace vessel and severely beat a Greenpeace crew member. The French government, exasperated by Greenpeace's campaign against its nuclear testing in the South Pacific, had secret agents blow up and sink the Greenpeace flagship, the *Rainbow Warrior,* a terroristic act that resulted in the murder of Greenpeace member Fernando Pereira. The U.S. Navy rammed a Greenpeace ship seeking to block a Trident submarine. Greenpeace director of toxics research Pat Costner's house was burned down by arsonists.

In addition to its practical achievements, Greenpeace is also highly significant as a model that demonstrates how to exploit the immense possibilities of television for radical change. Indeed, Greenpeace is arguably the first group working for social change, and certainly the first environmental group, whose *primary* rhetorical activity is the staging of image events for mass media dissemination.[2] Although media tactics are not new, Greenpeace is the first group both to explore fully and trust in the progressive potential of television, reflecting their Canadian lineage and the influence of Marshall McLuhan on key original members. For example, before joining Greenpeace in 1971, Hunter called McLuhan "our greatest prophet" (1971, p. 221). Paul Watson, an original member of Greenpeace and later founder of the Sea Shepherd Conservation Society, explains, "When we set up Greenpeace it was because we wanted a small group of action-oriented people who could get into the field and, using these McLuhanist principles (for attracting media attention), make an issue controversial and publicize it and get to the root of the problem" (quoted in Scarce, 1990, p. 101). Traditionally, radical activists on the left have been and continue to be wary if not contemptuous of mass media (e.g., Angus and Jhally, 1989b; McLaughlin, 1993) in favor of fetishizing immediacy (e.g., Baudrillard, 1981). This attitude is akin to what McLuhan describes as the "bulldog opacity" of literate people in response to the new technologies of mass media: "literate man[3] is not only numb and vague in the presence of film or photo [or video], but he intensifies his ineptness by a defensive arrogance and condescension to 'pop kulch' and 'mass entertainment'" (1964, p. 175). Early Greenpeace members took to heart McLuhan's aphorism "the medium is the message" and accepted McLuhan's challenge not to cower in their ivory towers bemoaning change but to plunge into the vortex of electric technology in order to understand it and dictate the new environment, to "turn ivory tower into control tower" (Hunter, 1971, p. 221). The early members of Greenpeace thought of themselves as media artists and revolutionaries, in line with McLuhan's contention that the "artist is the man in any field, scientific or humanistic, who grasps the implications of his actions and of new knowledge in his own time" (1964, p. 71).

In a book written shortly before Greenpeace's first image event, original member and early director Hunter argued that all revolutions are attempts to change the consciousness of the "enemy," and pointed out that in the past the "only medium through which a revolution could communicate itself was armed struggle." Today, however, the mass media provide a delivery system for strafing the population with mind bombs (Hunter, 1971, pp. 215–224). This philosophy of mass media has translated into a practice of staging image events based on the argument that "when you do an action it goes through the camera and into the minds of millions of people. The things that were previously out of mind now become commonplace. Therefore, you use the media as a weapon" (Hunter, quoted by Watson, in Scarce, 1990, p. 104). Fellow original Greenpeace member Watson elaborates, "The more dramatic you can make it, the more controversial it is, the more publicity you will get. . . . The drama translates into exposure. Then you tie the message into that exposure and fire it into the brains of millions of people in the process" (quoted in Scarce, 1990, p. 104).

Clearly, these early Greenpeace activists' theoretical insights on media could stand further development. They have a narrow conception of media that McLuhan would have frowned upon, they ignore a host of alternatives to armed struggle, and they adopt a causal model of media influence reminiscent of the discredited hypodermic needle model. Nonetheless, although theoretically a bit simplistic, in practice Greenpeace activists are sophisticated media artists who have been so successful that their artistry has been imitated by a legion of admiring radical environmental groups.

GUERRILLA IMAGEFARE IN THE WOODS

To protest logging on public lands in North Kalmiopsis, Oregon, home to "the most diverse coniferous forest on Earth" (Scarce, 1990, p. 67), Valerie Wade scales a yarder (a truck with a huge pole that uses cables to drag logs up and down steep slopes), perches precariously 90 feet up, and hangs a banner reading "From Heritage to Sawdust." To save old-growth forest, an Earth First! activist sits on a platform suspended 100 feet up in a giant Douglas fir, dwarfed by the trunk even at that height. Deep in the woods, a blue-capped, smiling, bearded head pokes up out of a logging road; the rest of the person is buried in the road. This attempt to stop logging by blockading the road adds new depth to the terms "passive resistance" and "active noncooperation." Such immobility, while making the tactic more effective, also renders the immobile activist more vulnerable to angry loggers and law enforcement officials.

On the 1981 spring equinox members of Earth First! unfurled a 300-foot-long plastic ribbon down the Glen Canyon Dam in order to simulate a crack in the dam, thus symbolically cracking this "monument to progress" clotting the Colorado River. With this image event (inspired by the Edward Abbey novel, *The Monkey Wrench Gang*), Earth First!, a radical, no-compromise environmental group founded a year earlier by five disgruntled mainstream environmentalists during a beer-besotted camping trip in the Pinacate Desert, debuted in the public consciousness. Since then, while Earth First! has deployed an array of tactics, most notably "ecotage" (ecological sabotage) or "monkeywrenching," in defense of natural ecosystems, image events have been their central rhetorical activity as they attempt to change the way people think about and act toward nature.[4]

In their efforts to put onto the public agenda issues such as clear-cutting of old-growth forests, overgrazing by cattle on public lands, de-pradations of oil and mineral companies on public lands, loss of biodi-versity, and the general ravaging of wilderness, Earth First! activists have resorted to sitting in trees, blockading roads with their bodies, chaining themselves to logging equipment, and dressing in animal cos-tumes at public hearings. Although these direct actions often fail in terms of accomplishing their immediate goals, their effectiveness as im-age events can be partially measured by the emergence of clear-cutting, old-growth forests, spotted owls, cattle grazing, and the 1872 mining law as hot-button political issues that national politicians are forced to re-spond to. For example, George Bush runs for office as the "environmen-tal president"; Al Gore is picked as the Democratic vice-presidential candidate in part because of his "green" credentials; President Clinton holds a "Forest Summit" in the Northwest over old-growth forest and the spotted owl; Congress of late perennially tries to reform the 1872 mining law; and the Clinton Administration initially advocates raising grazing fees on public lands (but later backs down in a move that signals to Clinton's political opposition that he can be browbeaten).

Earth First!, like Greenpeace before them, understands that the sig-nificance of direct actions is in their function as image events in the larg-er arena of public discourse. As philosopher and deep ecologist Bill De-vall explains, direct action "is aimed at a larger audience, and the action should always be interpreted by the activists. Smart and creative com-munication of the message is as important as the action itself" (quoted in Manes, 1990, p. 170). Although designed to flag media attention and generate publicity, image events are more than just a means of getting on television. They are crystallized philosophical fragments, mind bombs, that work to expand "the universe of thinkable thoughts" (Manes, 1990, p. 77).

Because Earth First!'s rhetoric and goals fundamentally challenge the discourse of industrialism and progress, their power is perhaps most evident in the vehemence of the counterrhetoric and backlash they have provoked. Newspapers have labeled Earth First!ers "tree slime," "human vermin," and the "eco-equivalent of neo-Nazi skinheads" (quoted in Short, 1991, p. 181). In the U.S. Congress Senator James McClure of Idaho compared Earth First! to "hostage-takers and kid-nappers" (Short, 1991, p. 185) and added a provision to an anti-drug bill making tree-spiking (a form of ecotage) a federal crime. At a 1989

Earth First! activists use a "tri-pod" and gate "lock-down" to block a logging road. Their signs read "NOT ONE MORE ANCIENT TREE" and "NO MAXXAM IN HUMBOLDT COUNTY."

campaign rally, Representative Ron Marlenee (R-MT) advised loggers to "spike an Earth First!er" (quoted in Lancaster, 1991, p. B1). The House, under cover of the Contract with America, passed legislation designed to gut environmental protection laws and deregulate industry (Helvarg, 1995a). Idaho Governor Cecil Andrus, who called Earth First! protestors "just a bunch of kooks," signed a "trespass" law that makes it a felony to interfere with logging activities, thereby equating nonviolent direct actions and civil disobedience (image events) with terrorism (Cockburn, 1995a, 1995b). The U.S. Forest Service has employed heavily armed "pot commandos" (law enforcement agents ostensibly cracking down on marijuana growers) to arrest protesters and has issued closure orders designed to prevent environmental activists from entering public lands where clear-cutting is going on, thus effectively preventing protests and silencing dissent. The FBI has used wiretaps and infiltrators in a $2 million surveillance operation against Earth First! known as Thermcon, which resulted in the arrests of a number of Earth First!ers, including cofounder Dave Foreman. As FBI infiltrator Michael Fain unwittingly revealed when he accidentally bugged his own conversation with two other agents, the arrests were political: "[Foreman] isn't really the guy we need to pop—I mean in terms of actual perpetrator. This is the guy we need to pop to send a message" (Manes, 1990, pp. 195–197). Evidently, even the FBI thinks in terms of image events.

Corporations used to exploiting resources on public lands with impunity have hired private investigators to spy on environmentalists and security firms to infiltrate Earth First! They have even ringed controversial logging sites on public lands with electronic motion detectors in order to monitor the movements of people in the forests, effectively transforming public lands into private security zones (Manes, 1990, p. 214). Also, corporations have filed strategic lawsuits against public participation (SLAPPs) against activists in efforts to silence them.

The backlash has also become physically violent, as workers, security personnel, law enforcement officials, and members of a corporate-sponsored grassroots anti-environmental movement known as Wise Use have attacked environmentalists (though Wise Use is against environmentalism in toto, they particularly target Earth First! and environmental justice activists). Besides being subjected to vandalism and death threats, hundreds of activists have suffered serious violence. Dave Foreman was run over at a blockade. Earth First!er Lisa Brown, who had locked her neck to a timber loader with a bicycle lock, was shot at by a security guard. Other activists have been beaten and tree-sitters have had the trees they were sitting in cut down. On September 17, 1998 David Chain, an Earth First!er, was crushed to death by a redwood when an irate Pacific Lumber logger continued felling trees despite the presence of Earth First! protesters (Goodell, 1999).

In 1988 Wise Use declared a "holy war against the new pagans who worship trees and sacrifice people" (Helvarg, 1994b, p. 648). Wise Use founder Ron Arnold declared, "We're out to kill the fuckers. We're simply trying to eliminate them. Our goal is to destroy environmentalism once and for all" (quoted in Helvarg, 1994a, p. 8). Since then there has been "an escalating campaign of beatings, arsons, pet killings, bombings, attempted murder, a rape and a possible homicide" (Helvarg, 1994b, p. 648). A dramatic example of this violent backlash is the 1990

Environmental activists risk physical harm from law enforcement officials and anti-environmental vigilantes. The above scene illustrates the violent backlash against Earth First!

bombing that maimed Judi Bari and injured Darryl Cherney, the two main organizers of Earth First!'s 1990 Redwood Summer campaign. Bari and Cherney were on their way to a rally when a pipe bomb under Bari's driver's seat exploded. Bari was seriously injured, her pelvis and coccyx shattered, and remained hospitalized for several months. Incredibly, Bari and Cherney were immediately arrested and charged with illegal possession of explosives. In attempting to frame the two, the FBI "told the Oakland Police the two 'qualified as terrorists'" (Rowell, 1996, p. 160)—this despite the lack of any record of violent activity by Bari and Cherney and their declared commitment to nonviolence. Indeed, Bari had just helped to write the "Redwood Summer Code of Non-Violence," which forbade verbal or physical violence and property damage. In addition, Bari and Cherney had received over 30 death threats. Furthermore, though they were accused of attempting to transport the bomb, "lab analysis by the FBI showed the bomb to be an anti-personnel bomb, found to contain a mechanism which would trigger an explosion if the vehicle in which it was placed was moved" (Rowell, 1996, p. 161). After harassing the radical environmental community for 2 months while investigating no other suspects, the FBI dropped the case. The bombing remains unsolved.[5]

CONTRARY IMAGES IN AMERICA'S HEARTLAND

Mourners trudge up the steps to the imposing Kentucky state capitol. The men, in particular, are dressed in the uniforms of ordinary folk: cowboy hats and overalls or baseball caps, jeans, and polyester shirts. Six pallbearers carry a wooden coffin draped with a Kentucky state flag. To the haunting strains of "Wayfaring Stranger," they lay the coffin down in the Rotunda. A woman in black delivers a eulogy, decrying that "the burial of Kentucky in waste can be witnessed continually and perpetually."

Those mourning the burial of Kentucky were members of Kentuckians for the Commonwealth (KFTC), a grassroots environmental justice group that achieved statewide prominence with a successful campaign calling for a constitutional amendment to severely curtail broad form deed strip mining, a practice whereby coal companies, waving a broad form deed granting the mineral rights to a piece of property allegedly signed, often with just an X by relatives 60 to 100 years ago, had the right to strip-mine the land, often leaving the surface property owners with uninhabitable land. Committed to a multi-issue approach, KFTC is presently working on the dumping of toxic and solid wastes, key issues in a region that "has recently become a sort of environmental Third World for more prosperous states that need to dump their trash somewhere" (Van Gelder, 1992, p. 62).

Among their tactics, KFTC affirms "direct-action organizing by members as KFTC's primary approach to change" (Zuercher, 1991, p. 136). To that end, KFTC set up a "lemonade stand" in the state capitol featuring samples of contaminated water collected in various eastern Kentucky communities; put up "wanted posters" of a state official who was avoiding them; baked and sold "Cookies for Cleanup"; offered state officials regular, decaffeinated, leaded, or unleaded coffee at a public hearing on water contamination; wore decontamination suits to the Natural Resources and Environmental Protection Cabinet offices, acted out a live political cartoon depicting state legislators in bed with industry lobbyists; and held the previously described funeral for Kentucky (Zuercher, 1991, pp. 121, 151—152; Van Gelder, 1992, pp. 62–67).

While public officials and even members of the media have condemned such tactical image events as "gimmicks" and "grandstanding," Zuercher (1991) notes that "KFTC's creative tactics also receive wide media coverage and get results" (p. 152). For example, the funeral for Kentucky was held when the legislature was not in session, so, as KFTC member Lisa Abbott explains, "There was no audience but great reports. The news coverage was very successful" (personal communication, March 14, 1995). Indeed, from KFTC's perspective, it is "through creative, direct, symbolic and even daring actions" that they have "affected the balance of power" (Zuercher, 1991, p. 151).

> On a gray day spitting snow, a helicopter circles a bridge. Between the supports of the bridge flies an American flag. Beneath the flag hangs a banner that reads "Grandparents for the Future." On the bridge itself, surrounded by several dozen supporters wearing yellow "Allegany: No Dump" masks, sit six elderly people bundled up in lawn chairs and handcuffed to a chain traversing the bridge. Police and state troopers, after wading through news media, cite the protesters (one is 87 years old) for violating a New York State Supreme Court preliminary injunction forbidding citizens from phys-

Grandparents and members of the Allegany County Non-violent Action Group (ACNag) occupy a bridge in order to block a commission seeking a site for a radioactive waste dump. To secure their place, they are chained to the bridge. In the end, ACNag was successful and no radioactive waste dump was sited in Allegany County or anywhere else in New York.

ically interfering with the activities of the Siting Commission, which is seeking a spot for a low-level radioactive waste dump. The grandparents are also cited for disorderly conduct. As they are being arrested and are asked to identify themselves, each responds in turn, "Allegany County."[6]

———

Roland Warren, the brains behind the bridge blockade and one of the bridge-sitters, explains the rationale behind the image event: "It would be very symbolic. People around here respect older people. And, of course, it seemed to me it would be embarrassing, as it turned out to be, for the state police to come in and arrest us" (O'Shaughnessy, 1993).

In their struggle to prevent the siting of a low-level radioactive waste dump in their rural county, the ACNag employed various tactics to block the Siting Commission. In the penultimate confrontation, ACNag activists on horseback confronted the Siting Commission. The New York State Police, feeling threatened by the huge, skittish beasts, responded with excessive force. The police tore one rider off his horse, surrounded him, and then roughly arrested him. The confrontation made the *New York Times,* and the next day Governor Cuomo ordered the Siting Commission to cease trying to site a low-level radioactive waste dump. Instead, the state sued the federal government over the law requiring states to take responsibility for low-level radioactive waste generated within their borders. On June 19, 1992, the U.S. Supreme Court sided with the states. On May 11, 1993, the Siting Commission voted to remove Allegany County from the preferred list of sites (O'Shaughnessy, 1993).

KFTC and ACNag are both part of the proliferation of environmental justice groups as local citizens struggle to protect their habitats from the hazards of waste incineration, dumping, and toxic industries. The groups conceptualize their struggle as being about environmental justice because their communities are often targeted as sites due to class discrimination, institutional racism, and regional bias. These local grassroots groups are loosely linked by regional and national organizations, such as the Citizens' Clearinghouse for Hazardous Wastes (CCHW),[7] which serve as resource and information centers. Such linkages enable environmental justice groups to articulate their struggle as NIABY ("not in anyone's backyard") instead of NIMBY ("not in my backyard"). ACNag, for example, has sent members to Nebraska, North Carolina, and South Carolina to share tactics with groups in those states fighting low-level radioactive waste dumps. The goal, then, becomes "stopping the toilet" (Montague, 1993, p. 15) of the industrial system by making waste disposal difficult and expensive.

Employing image events and other tactics, these groups have been incredibly effective: "there has not been a single new hazardous waste site opened in the last ten years. Without passing any new laws or regulations . . . we have stopped the expansion of hazardous-waste sites in this country" (CCHW founder Lois Gibbs quoted in Greider, 1992, p. 169).[8] As Peter Montague, executive director of the Environmental Research Foundation, explains, deploying the tactic of image events to stop up the industrial toilet works well:

> For example, over the past seven years, the disposal of so-called "low-level" radioactive waste became increasingly difficult and expensive. As a direct result, the production of low-level radioactive wastes declined from 2.7 million cubic feet per year in 1985 to 1.4 million cubic feet per year in 1991—a 48 percent reduction in six years. Reporting this news, the *New York Times* gave the reason: "The increasing problem with disposal has had some benefit. . . . Scientists are turning to methods that don't involve radiation. Companies and hospitals have become more efficient in using radioactivity and now often clean and reuse gloves and other equipment rather than discarding them." (1993, p. 15)

The successes of environmental justice groups prompted President Clinton to highlight environmental justice as a campaign issue. On the last day of his train trip to the Democratic National Convention, Clinton declared:

> We cannot go forward together as a country, a country where it works for all of us, unless we have a shared commitment to protect the environment and unless we want to protect everybody's environment. . . . Today I am calling for a new national commitment to help protect all communities from toxics by the year 2000. First, I am determined that finally we clean up the toxic waste sites that scar our landscape and threaten our neighborhoods. (1996, p. B11)

Although Clinton's announced support should not be mistaken for real commitment, it does point to the success of environmental justice groups in gaining visibility and public support for their issues.

Like members of other environmental groups, environmental justice activists have been the victims of a violent backlash, evidence of both their success and the stakes involved in their struggle. Lois Gibbs, founder of the Love Canal Home Owners Association and later founder of CCHW, notes, "People have been followed in their cars, investigated by private detectives, had their homes broken into. I'd say 40 percent of people protesting toxic waste sites and incinerators around the country have been intimidated" (Helvarg, 1994b, p. 651). Antitoxics activist Paula Siemers has been knocked unconscious, stabbed, had her dog poisoned, and had her house set on fire. Stephanie McGuire, an activist against water pollution, was raped and tortured by three men in camouflage uniforms. "After they cut my throat, they poured water in it from the river and said 'Now you'll have something to sue about'" (McGuire, quoted by Helvarg, 1994a, p. 374; see also *60 Minutes*, 1993).

RHETORIC, REASON, AND IMAGE EVENTS IN THE PUBLIC SPHERE

The preceding snapshots of the image events of four environmental groups, besides sharing the "stuff" upon which I am meditating, were also designed to argue for the unorthodox tactic of staging image events as the primary rhetorical activity of environmental groups that are radically challenging and even changing public consciousness in the United States (if not the entire industrialized world). When taken seriously as rhetorical activity, image events challenge a number of tenets of traditional rhetorical theory and criticism, starting with the notion that rhetoric ideally is "reasoned discourse," with "reasoned" connoting "civil" or "rational" and "discourse" connoting "words." Although there have been challenges to such a narrow notion of rhetoric, in practice this conventional conceptualization remains prevalent.

For example, in their popular *Contemporary Perspectives on Rhetoric*, Foss, Foss, and Trapp "define rhetoric broadly as the uniquely human ability to use symbols to communicate with one another" (1985, p. 11). On the very next page they argue that "the paradigm case of rhetoric is the use of the spoken word to persuade an audience. Examples would be a lawyer arguing before a jury, a legislator attempting to persuade the legislative body to pass a major bill, a minister addressing a congregation, or a politician attempting to persuade the populace to vote for her" (p. 12). Rhetoric is still conceived as civil, reasoned, verbal discourse.

Such a traditional and yet still contemporary perspective discounts image events as rhetoric and marginalizes the groups that practice such a form of rhetoric. Liberal political columnist William Greider despairingly calls such tactics "the politics of rude and crude" that reveals "the disconnectedness that prevents them [citizens] from entering into any kind of enduring, responsible relationship with those in power" (1992, p. 163). To dismiss image events as rude and crude is to cling to "presuppositions of civility and rationality underlying the old rhetoric," a rhetoric that supports those in positions of authority and thus allows civility and decorum to serve as masks for the protection of privilege and the silencing of protest (Scott and Smith, 1969, pp. 7, 8). Indeed, Aristotle's (1991) *On Rhetoric: A Theory of Civic Discourse* can be read as a primer on how to maintain hegemony. The field, in general, has followed Aristotle's example. As Scott and Smith observe, "Since the time of Aristotle, academic rhetorics have been for the most part instruments of established society, presupposing the 'goods' of order, civility, reason, decorum, and civil or theocratic law" (1969, p. 7). Such an understanding of rhetoric assumes a consensus on fundamental values and a belief in the system, which is antithetical to the very purpose of groups that are trying to produce social movement by challenging the legitimacy of the establishment. As Lois Gibbs of CCHW explains, "The movement is outside the system" (quoted in Greider, 1992, p. 168), and thus must rely on rhetoric that challenges the system's values and authority.

The social protests of the 1960s and early 1970s forced scholars to consider the implications for rhetoric of extralinguistic confrontational activities. However, as Brant Short points out, "Although critics acknowledged the rhetorical aspects of confrontation, protest, and agitation, these studies suggest that theoretical accounts of seemingly *nonrational* discourse remained linked to traditional notions of logic, rationality, and artistic proofs" (1991, p. 173). So confrontational rhetoric was measured against the ideals of reasoned discourse (Haiman, 1967) or was seen as a technique to gain attention for the "real" rhetoric (McEdwards, 1968). Even when a few scholars started to study agitation, coercion, and confrontation as forms of rhetoric in their own right (Scott and Smith, 1969; Bowers and Ochs, 1971; Simons, 1972), their traditional definitions of rhetoric reduced agitation, coercion, and confrontation to instrumental activity, not constitutive action.

There have been notable exceptions to the tendency to reduce confrontational rhetoric to an instrumental activity. In 1971 Gregg suggested that social protest rhetoric primarily serves the ego function of self-affirmation for the protesters. Along similar lines, a number of scholars have studied how social protest rhetoric constitutes an identity for the protesters. Karlyn Kohrs Campbell (1971) explored how the rhetoric of radical black nationalism worked to create a new black "people." Windt (1972) argued that the diatribes and antics of the Cynics and Yippies served not merely to attract attention but also to mark their identities and confirm their beliefs. In his analysis of Red Power, Lake (1983) found that the purpose of Native American protest rhetoric was not the instrumental one of influencing whites but the constitutive one of reconstituting the traditional Native American and their ways of life. All of these studies understand linguistic and extralinguistic social protest rhetoric as working to constitute the identities of protest groups opposed to or outside of the dominant culture. Although continuing to understand such protest rhetoric as constitutive, my analysis of the image events of radical environmental groups shifts the focus of attention from how such unorthodox rhetoric constitutes the identities of protest groups to how it reconstitutes the identity of the dominant culture by challenging and transforming mainstream society's key discourses and ideographs.

In general, old habits die hard. When the more obvious forms of agitation and confrontation of the 1960s and early 1970s ceased, the call to "open ourselves to the fundamental meaning of radical confrontation" (Scott and Smith, 1969, p. 8) lost much of its force. Some heralded with glee the turn away from confrontation and social movements and the return to the reasoned discourse of traditional rhetoric: "Times change, and with them the goals, values, and orientations of scholars. While the Age of Aquarius was preoccupied with the rhetoric of the streets, the Age of Reagan has restored the rhetoric of the platform. Studies of the New Left are passé, scholarship in the rhetoric of social movements is moribund, and it is hard to find even a glimmer of interest in confrontation as a rhetorical strategy" (Lucas, 1988, p. 243). Others turned away from a narrow sense of confrontation and have argued for an expanded sense of social movement that highlights strategies of identification, accommodation, affirmation, and enactment (Henry and Jensen, 1991; Sheedy, n.d. For examples, see Carlson, 1986; M. R. Williams, 1994; Powell, 1992, 1995).

Even for its proponents, the power of the call for rhetoricians to "open ourselves to the fundamental meaning of radical confrontation" waned. In 1969 Robert Scott pushed a rhetoric of confrontation that involved "killing the enemy" (Scott and Smith, 1969, p. 5) and chastised rhetoricians as serving the establishment interests of the haves. Later, though, in the influential text *Methods of Rhetorical Criticism,* Scott and his coauthors define rhetoric as "the human effort to induce cooperation through the use of symbols" (Brock, Scott, and Chesebro, 1989, p. 14). Simons, who in 1972 criticized the tendency since Aristotle onward to view rhetorical practice "as a process of finding *common cause* with one's hearers" (p. 236, emphasis in original) and condemned the predominant "rhetoric for insiders" (p. 236) as inapplicable to studying social conflict and groups agitating for social change, spent the rest of the 1970s, 1980s, and 1990s elaborating a sociological theory of social movements that focuses on organizational rhetoric and equates social movements with corporations and their leaders with CEOs (1970, 1980, 1991; Simons, Mechling, and Schreier, 1984).

Even Short, who is explicitly critical of the tendencies to tie nonrational discourse to traditional tenets of rhetoric and reduce it to mere attention-getting devices (1991, p. 173), reiterates these tendencies later in that same essay. After analyzing the confrontational rhetoric of Earth First!, Short concludes that their confrontational rhetoric promotes environmentalism by "draw[ing] public attention to many concerns of the larger environmental movement" and by "push[ing] mainstream environmental groups to respond to controversial issues" (1991, p. 183). Once again, rhetorical analysis has reduced groups outside the mainstream and their unorthodox rhetorical tactics to the entirely ancillary roles of gadflies and attention-seekers.

In this study, then, I accept Scott and Smith's challenge to be open to the fundamental meaning of radical confrontation and thus neither to dismiss image events as gimmicks or the antics of the unruly nor to reduce them to flares sent out to gain attention for the "real" rhetoric. Working from an understanding of rhetoric as the mobilization of signs for the articulation of identities, ideologies, consciousnesses, communities, publics, and cultures,[9] I am exploring how radical environmental groups are using image events to attempt both to deconstruct and articulate identities, ideologies, consciousnesses, communities, publics, and cultures in our modern industrial civilization.[10]

Taking image events seriously also challenges the association of rhetoric with a notion of discourse as limited to words. Although rhetorical theorists from Aristotle through Bacon to Perelman have recognized the importance and power of "bringing-before-the-eyes," "making pictures," and "creating presence," today "in the age of television, dramatic, digestive, visual moments are replacing memorable words" (Jamieson, 1988, p. x). Kathleen Hall Jamieson even argues that "speech in such settings would dilute the power of the nonverbal message being telegraphed to audiences, regardless of their native language, throughout the world" (1988, p. 115). Thus, far from being the desperate stunts of the disillusioned, image events are the central mode of public discourse both for conventional electoral politics (Jamieson, 1988; Gronbeck, 1992, 1995; Postman, 1985) and for alternative grassroots politics in an era dominated by a commercial televisual electronic public sphere. As Gronbeck bluntly puts it, "The telespectacle [image event], for better or worse, is the center of public politics, of the public sphere. . . . We must recognize that the conversation of the culture is centered not in the *New York Review of Books* but in the television experience" (1995, p. 235).[11] And while the civil rights movement of the 1950s and 1960s was catalyzed by powerful pictures *and* eloquent words, radical environmental groups rely almost solely on image events to create social movement. Indeed, it is telling that there are no famous environmental speakers and no memorable environmental speeches, except perhaps for Chief Seattle's letter to the president of the United States, which was fabricated by a white man in the 1970s (Egan, 1992).[12]

Though there has been some work on a rhetoric of images,[13] too often rhetorical scholars keep returning to talk even when ostensibly looking at television. This approach not only blinds them to what is significant about television, it allows them to retain their old terms and models, thus sacrificing insight for habit. For example, in "The Enthymeme as Postmodern Argument Form," Roger Aden posits that we live in a postmodern age in large part due to the emergence of television and the consequent proliferation and endless circulation of signs (1994, p. 54). Yet in his study Aden focuses exclusively on the *words* of David Duke on *The Phil Donahue Show* (the choice of a talk show is telling) in order to claim that the paradigmatic form of public argument in our postmodern television age is the enthymeme: "Ironically, today's postmodern audiences process arguments in a manner eerily similar to the classical audiences of ancient Greece. . . . Ironically, this form of argument processing mirrors that posited by Aristotle in his explanation of the deductive reasoning form he labeled an enthymeme" (1994, pp. 54, 55).

Although Aden's essay is interesting, the only way to examine television as a public sphere and end up rediscovering the Greek Pnyx is to ignore images. My point here is not to criticize Aden, but rather to point to his essay as symptomatic of a tendency in the discipline of rhetoric to study television and other imagistic media by focusing on words to the neglect of images. For example, in an important essay that won the 1995 Golden Anniversary Monograph Award and marks a significant development in G. Thomas Goodnight's conceptualization of the public sphere, Kathryn Olson and Goodnight explore the image events of anti-fur activists as oppositional arguments that "presage the advent of a contemporary public sphere" (1994, p. 253). Similar to Aden's move, they immediately place the image events within the traditional discursive frame of enthymemes, with the image events functioning to block enthymematic associations. Although Olson and Goodnight recognize that image events "entwine discursive and nondiscursive argumentation" (1994, p. 251), their analysis implicitly instantiates a hierarchy that privileges the discursive (words) over the nondiscursive (images).[14]

They do this in two ways. First, they discuss discursive arguments before discussing nondiscursive arguments (1994, pp. 251–252, 257, 260). Second and more significant, when analyzing nondiscursive arguments (advertisements, posters, direct confrontations, performances), Olson and Goodnight consistently emphasize the words and neglect the images, providing no close readings of the images. Images and image events are reduced to their captions. For example, a protest by Friends of

Animals in front of the Fifth Avenue Fur Vault is described as follows: "a banner reading 'Peace on Earth Begins with Your Wardrobe' and pictures of animals captioned 'Let me Live'" (1994, p. 257). Anti-fur advertisements do not inspire Olson and Goodnight to more elaborate readings of images: "some ads display endearing pictures of live animals and bear captions like 'Meet one of the 65 pelts it takes to make a single fur coat' and 'When you choose to wear fur, animals suffer and die needlessly'" (1994, p. 260). This pattern of analyzing images is repeated throughout the essay, so that "close" readings of images consist of quoting captions and sometimes throwing in adjectives like "graphic," "grisly," "colored," and "dead" (1994, pp. 258, 261, 263).

It is striking that an essay that is overtly sympathetic to the rhetorical study of images can be so neglectful and noncomprehending of the images it analyzes. Olson and Goodnight's bias toward the traditional discursive parameters of rhetoric surfaces revealingly in their concluding sentence, wherein they praise image events as "rich moments of rhetorical invention" while still hoping that "genuine deliberation may emerge through social controversy" (1994, p. 273).

As in Aden's case, my criticism of Olson and Goodnight is a criticism of a disciplinary tendency. For still another example, see Michael Hogan's provocative book on the nuclear freeze movement, wherein he condemns the nuclear freeze campaign as a "telepolitical movement" that eschewed ideas and instead "waged a war of images, slogans, and bandwagon appeals" (1994, p. 7). Although recognizing and decrying the centrality of "good visuals" to political discourse in a televisual public sphere, Hogan analyzes what he terms "the televisual freeze movement" without recourse to studying the televised images. Instead, Hogan relies on the words in the transcripts of the televised broadcasts to interpret "the televisual freeze movement" (see esp. pp. 141–167). In short, Hogan performs a rhetorical criticism of a televisual social movement dependent on good visuals while neglecting the televised images.

Although it is possible to study television and other image-dominated media by focusing on words, why bother? Though scholars often study it as such,[15] television is not radio with pictures, and the meaning of images is not captured by captions. To think such is to miss everything important about imagistic discourse. To understand the rhetorical force of the televisual/imagistic public sphere requires a "reading" of images that resists using our ready-to-hand theoretical tools, or at least resists using them in familiar ways.

My support of environmental groups' adaptation of image events as necessary tactics is not meant to suggest that television is a level playing field, Habermas' idealized bourgeois public sphere of undistorted communication. Obviously, elected officials, especially the president and members of Congress, as well as business leaders and corporations, enjoy an enormous advantage over environmental groups in terms of access to media, particularly television, and control of their image, due in no small measure to the fact that the media are themselves giant corporations with a vested interest in the status quo (see Donovan and Scherer, 1992; Moyers, 1989; Herman and Chomsky, 1988; Parenti, 1993). The implications of these disadvantages will be explored in Chapter 5. Still, in order to participate in the most important arena of public discourse, the televisual public sphere, and in order to be more than an enclave, environmental groups must use the tactic of image events. The distinction here is one of strategy versus tactic. Those in positions of institutional power are able to use image events in strategic ways, but for radical environmental groups the use of image events is a tactic, "a maneuver 'within the enemy's field of vision' . . . and within enemy territory," a recognition that they "must play on and with a terrain imposed on it and organized by the law of a foreign power" (de Certeau, 1988, pp. 36–37). The image events of environmental groups are tactics that operate in the territory of the system but outside the sense-making rules or the lines on the grid of intelligibility of the system—a necessary condition if they are to create social movement.

I need to explain what I mean by the term "public sphere." Within a conventional usage, Habermas' liberal public sphere, image events do not register—that is, they neither count nor make sense within the rules, the formal procedures, of such a public sphere. They do not consist of talk by preconstituted rational subjects directed towards consensus, the deliberative rhetoric that, according to Goodnight, is characteristic, even constitutive, of the public sphere (1982, pp. 214–215). Indeed, Habermas would likely point to image events as further evidence of the disintegration or refeudalization of the public sphere—a return to the spectacle of the Middle Ages. Although the image events of radical environmental groups are often spectacular, they are not the displays of the rulers but, rather, the discourse of subaltern counterpublics (Fraser, 1992, p. 123) who have purposely been excluded for political reasons from the forums of the public sphere by the rules of reason and the protocols of decorum. If the bourgeois public sphere was originally an arena for publicity by which the citizens held the state accountable, today, in an electronic age, environmental activists (as well as other citizen activists) are deploying image events to generate publicity in a televisual public sphere in order to hold the state and corporations accountable. From this perspective, then, while Habermas' conceptualization and history of the public sphere has many flaws (see Curran, 1991; Dahlgren and Sparks, 1991; Eley, 1992; Fraser, 1992; Landes, 1988; McLaughlin, 1993; Pateman, 1988; Peters, 1993; and Ryan, 1992), foremost among them his fetishization of a procedural rationality at the heart of the public sphere, which serves as an exclusionary and impoverished normative ideal that shuns much of the richness and turbulence of the sense-making process, still, as Nancy Fraser argues, "something like Habermas' idea of the public sphere is indispensable to critical social theory and democratic political practice" (1992, p. 111).

The concept of the public sphere is indispensable for theoretical and practical reasons. The public sphere is a compelling spatial metaphor[16] that has captured the imagination of social theorists across disciplines. As a conceptual tool, it is a particularly apt lever for getting at the dilemmas of democracy in an industrial/technological age of mass communication and oligarchy. Goodnight's successful importation of the term into rhetorical circles (1982, 1987) has both sparked and circumscribed debate (Hauser, 1987; Balthrop, 1989; Biesecker, 1989a; Birdsell, 1989; Hynes, 1989; Olson and Goodnight, 1989; Schiappa, 1989; Wallinger, 1989). For a critical rhetorician not to engage and contest Goodnight's Habermassian public sphere is to surrender an important part of the field of rhetorical theory, with serious consequences for thinking about rhetoric, politics, radical democracy, and citizenship, consequences that implicate critical rhetorical practices.[17]

In a social field characterized more by the conflictual process of hegemony than by communal deliberation and community consensus, radical environmental groups are competing in a corporate-owned public sphere that needs to be understood not as a civic forum but as "the structured setting where cultural and ideological contest or negotiation among a variety of publics takes place" (Eley, 1992, p. 306). This conceptualization of the public sphere avoids restricting it to the medium of talk characterized by rationality and recognizes that "the public sphere was always constituted by conflict" (Eley, 1992, p. 306).

In other words, although today's televisual public sphere is not the liberal public sphere of which Habermas dreams, wherein a reasonable public through deliberative discussion (note the congruence with rhetoric's emphasis on reasoned discourse) achieves a rational public opinion (a dream that "made possible the democratic control of state activities" [Habermas, 1974, p. 50] and still "grounds the constitutional state normatively" [Peters, 1993, p. 544; see also Habermas, 1974, pp. 52–53]), neither is it the medieval public sphere of representative publicity that Habermas fears, a site where rulers stage their status in the form of spectacles before the ruled. Rather, in today's televisual public sphere corporations and states (in the persons/bodies of politicians) stage spectacles (advertising and photo ops) certifying their status before the people/public *and* subaltern counterpublics participate through the performance of image events, employing the consequent publicity as a social medium through which to hold corporations and states account-able, help form public opinion, and constitute their own identities as subaltern counterpublics. Critique through spectacle, not critique versus spectacle.

Heeding metaphorically Husserl's call for a return to the things-in-themselves (metaphorical in that we never have access to things-in-themselves outside of any discourses), I have started this study with a general look at image events. In some respects this approach was neces-sary because traditional frames of rhetoric and politics render image events invisible or frivolous. In trying to make sense of image events, then, I have had to rethink rhetorical theory through the practice of im-age events, which has resulted in the questioning of certain rhetorical paradigms, foundations, and truths and the opening of possibilities for rhetorical theory, critical rhetoric, and politics. This book is an account of those meditations.

MEDITATION I

Doubtless it is more necessary . . . to transform concepts, to displace them, to turn them against their presuppositions, to reinscribe them in other chains, and little by little to modify the terrain of our work and thereby produce new configurations; I do not believe in decisive ruptures, in an unequivocal "epistemological break," as it is called today. Breaks are always, and fatally, reinscribed in an old cloth that must continually, interminably be undone.

—Jacques Derrida, *Positions*

CHAPTER 2

THE RHETORIC
OF SOCIAL MOVEMENTS
A Theoretical Diagnostics and Overhaul

Scholars have tended to study the amorphous concept of social movements from an organizational perspective, even within rhetoric circles. A rhetorical perspective, however, offers unique advantages. This is particularly true today as mass media render organizational size and resources largely irrelevant and as the "new social movements" (environmental, women's, civil rights, gay) challenge earlier sociological conceptions of social movements. Such a rhetorical, or at least discursive, turn is evident among sociologists, who once focused on organizations and resources when trying to make sense of social movements (the resource-mobilization approach). Recently, particularly in Europe, those sociologists have been groping their way toward rhetoric as they struggle to make sense of the new social movements.[1]

Scholars need to focus on the new wave of social mvmts.

As a number of sociologists (J. L. Cohen, Melucci, Offe, Touraine) note, the new social movements differ from past social movements in two fundamental ways: issues and organizations. For a variety of reasons, the new social movements do not focus on the distribution of material goods, the expansion of institutional political rights, and security, but rather thematize personal and collective identity, contest social norms, challenge the logic governing the system, and, in sum, deconstruct the established naming of the world. In other words, this is a shift from economic grounds to cultural grounds or from the domains of the state and the economy to the domain of civil society, in part "because the domination which is challenged controls not only 'means of production' but the production of symbolic goods, that is, of information and images of culture itself'' (Touraine, 1985, p. 774; see also Melucci, 1985, pp. 795–796). Organizationally, new social movements resist formal modes of organization (unions or parties) and instead create networks of grassroots groups lacking hierarchy and much in the way of resources.

These characteristics of new social movements have led many sociologists to abandon the traditional and resource-mobilization paradigms and even to question the concept of social movement. These sociologists, grouped under the identity-oriented paradigm,[2] now argue for the need "to stop treating it [social movement] just as an empirical phenomenon" (Melucci, 1985, p. 793) because the notion of social movement "does not describe part of 'reality' but is an element of a specific mode of constructing social reality" (Touraine, 1985, p. 749). This perspective leads to a definition of social movements as "action systems," "forms of collective action" (Melucci, 1985, pp. 792, 795), or "conflicts around the social control of the main cultural patterns" (Touraine, 1985, p. 760). There is, then, a clear move in their definitions from social movement as an object to social movement as an activity. This represents a radical break for sociologists, leading Touraine to argue that sociology must "separate itself from an old definition of its object as the *study of society*, which should be replaced by the *study of social action*" (1985, p. 782).

This break also marks a significant though incomplete turn toward a rhetorical theory of social movements. These sociologists recognize as a crucial form of social action "the processes of communication engaged in by contemporary collective actors as they articulate new identities and societal projects" (J. L. Cohen, 1985, p. 704). This social action is rhetoric, though its name remains unspoken, perhaps due to the European heritage of the identity-oriented sociologists.

Missing rhetoric, they stumble. Touraine notes that the term "social movements" can be replaced by "public opinion transformations" (1985, p. 786), but goes no further. Disappointingly, Melucci, who adamantly argues that a social movement is not a "thing" but instead a "symbolic challenge to the system" that affects "collective consciousness" (1985, pp. 792, 812, 814), concludes that the success or failure of symbolic challenges cannot be measured (1985, p. 813) and so retreats to characterizing social movements as "social organizations" (1985, p. 813). If Melucci had approached these symbolic challenges from a rhetorical perspective, he might have considered language strategies to be as material as organizational patterns, measurable through the analysis of the use of key terms in public discourse: ideographs, images, metaphors, narratives. Cohen, who is attempting the difficult project of producing a synthesis of the resource-mobilization and identity-oriented paradigms, recognizes that the "access of interpretation to identity is through the interrogation of forms of consciousness" (1985, p. 665) and suggests examining the theories (ideologies) participants produce for and within movements. In other words, Cohen is suggesting that theorists must examine the rhetoric of activists attempting to produce social change. Unfortunately, without a notion of rhetoric, Cohen turns to Habermas in an unsatisfactory conclusion.

WHAT ARE SOCIAL MOVEMENTS?

Ironically enough, while the last two decades have witnessed a slow turn in sociology to a discursive if not rhetorical vision of social movements, rhetorical theory has imported a traditional sociological approach that emphasizes organizations and resources at the expense of rhetoric. Through a complex interaction of disciplinary, institutional, and perhaps interpersonal relations, since the early 1970s Herbert Simons' sociological approach to social movement has been the dominant paradigm in the discipline of rhetoric. This dominance is both evidenced and reinforced by Simons' pride of place in three important publications, the special movement issue in 1980 of the *Central States Speech Journal*, Arnold and Bowers' (1984) weighty *Handbook of Rhetorical and Communication Theory*, and the special issue on social movement criticism in 1991 of *Communication Studies*. From his privileged perch as commentator with the last word in the two journal issues and as reviewer of record (along with coauthors Mechling and Schreir) of "The Rhetoric of Social Movements" in Arnold and Bowers' definitive tome, Simons surveys his theoretical domain, praising the faithful (comments on Stewart in 1991, p. 95), correcting the wayward (on Darsey in 1991, pp. 97–98), excommunicating the heretics (on McGee and Zarefsky in 1980, 1991; on Condit and Lucaites in 1991, pp. 96–97), and continually redefining the boundaries while simultaneously policing them (on Nelson, 1991, pp. 98–101). The end result is that Simons' hegemonic rhetorical theory of social movement is a disciplinary achievement that renders invisible many groups and tactics. Quite simply, within the grid of intelligibility that structures Simons' domain, radical environmental groups and their image events do not register. They are unintelligible and therefore do not count.

Early on, in "Requirements, Problems, and Strategies," Simons emphasizes the importance of theory for practice: "By suggesting parameters and directions to the rhetorical critic, theory places him in a better position to bring his own sensitivity and imagination to bear on analyses of particular movements" (1970, p. 2). In a move away from Leland Griffin's historical (1952, 1964) and dramatistic (1969) orientations toward social movements, Simons chooses a "leader-centered conception of persuasion in social movements" rooted in sociological theory (1970, p. 2). Simons explicitly defines a social movement "as an uninstitutionalized collectivity that mobilizes for action to implement a program for the reconstitution of social norms or values . . . from the bottom up" (1970, p. 3; Simons et al., 1984, p. 792). Several important consequences follow from Simons' theoretical choices. Simons hints at some of these consequences when he writes that "the rhetoric of a movement must *follow*, in a general way, from the very nature of social movements" (1970, p. 2). Thus, Simons' focus of study is not so much rhetoric as a particular type of collectivity that he terms a "social movement," since the rhetoric is determined by the nature of the social movement. From this perspective, rhetoric is reduced to an instrument for fulfilling functions. In short, Simons' approach is not rhetoric-centered but rather has "a focus on the people and events that 'cause' social changes, relegating public discourse to a supporting role in the story" (Condit, quoted by Henry and Jensen, 1991, p. 86). The tendency in such an approach is to reduce rhetoric to its instrumental use by leaders in what is defined as a social movement.

Defining a social movement as a collectivity is also problematic, especially for a rhetorical theorist. *Webster's Encyclopedic Unabridged Dictionary* offers three definitions of *movement* that apply to social movement: (1) a progressive development of ideas toward a particular conclusion; (2) a series of actions or activities directed or tending toward a particular end; and (3) a diffusely organized or heterogeneous group of people or organizations tending toward or favoring a generalized common goal (1989, p. 936). The first two definitions seem particularly amenable to a rhetorical theory of social movement. Simons chooses the third.

McGee mocks this approach as studying movement only by a stretch of the imagination (1975, p. 235; 1980c). For McGee, "social movement is a set of meanings and not a phenomenon" (McGee, 1980c, p. 234), and a rhetorical theory of movement must be a hermeneutical account of human consciousness, not an account of human organizational behavior (McGee, 1980c, p. 242). McGee argues that a social movement as a phenomenon, an empirical object, does not exist. As evidence, he points to the impossibility of defining such an object. Such a charge is largely supported by the work of sociologists and even by Simons himself. For example, a recent article, "The Ecological Movements in the Light of Social Movements' Development," in the *International Journal of Comparative Sociology*, in trying to define social movement, notes that "the social movements' scope is wide and the criteria remain elusive" (Frechet and Worndl, 1993, p. 57). Stewart, Smith, and Denton, in *Persuasion and Social Movements,* write that "attempts to define the term 'social movement' have added to rather than lessened the confusion and disagreement" (1989, p. 4). Contrary to Simons' claim in response to McGee that there exists "near agreement among sociologists and others on the term 'social movement'" (1980, p. 308), actually, as sociologist Cohen concludes, "there is little agreement among theorists in the field as to just what a *movement* is" (1985, p. 663; see also Melucci, 1980, p. 199, and 1984, p. 823).

This confusion is clearly evident in the special issue of *Communication Studies* (1991, Vol. 42, No. 1) on social movement criticism. In his closing commentary on the articles in the issue, Simons, in commenting on a piece about Mussolini, admits to the existence of top-down institutionalized social movements, and cites as examples Mussolini, Hitler, Gorbachev, Carter, and Reagan, among others (1991, pp. 98–99). This change renders Simons' definition of social movements inoperable: "noninstitutionalized [or institutionalized] collectivities that promote or resist social change from the bottom-up [or top down]."

To compound his troubling choice of definition, Simons throughout the "Requirements" article equates the collectivity he terms a social movement with an organization (1970, pp. 2, 3, 4, 11), so that a social movement is like a corporation except for the differences resulting from being an informal organization (thus the civil rights movement is similar to IBM). This analogy is extended to the leaders of social movements, so that they are like CEOs: "Like the heads of private corporations or government agencies, the leaders of social movements must meet a number of rhetorical requirements" (Simons, 1970, p. 3). Simons' perspective forces him to focus on the internal rhetoric of a social movement required to make the social movement internally coherent and effective in the larger corporate world. In the end, Simons' rhetoric of social movements is equivalent to organizational communication or, in the words of Cathcart, managerial rhetoric (1980, p. 272).

This reduction is foretold by Simons himself when he argues that "the standard tools of rhetorical criticism are ill-suited for unraveling the complexity of discourse in social movements or capturing its grand flow" (1970, p. 2), and consequently delimits the scope of his inquiry to exclude general social movements (Simons et al., 1984, p. 793). In essence, at the inception of his work on rhetoric and social movement, Simons admits that he is incapable of understanding social movement from a rhetorical perspective. Although such an admission explains Simons' turn to sociology, it does little to justify his preeminence as the dean of rhetoric and social movement studies. As McGee correctly if somewhat harshly concludes: "Simons et al. do not *require* either the concept 'movement' or the concept 'rhetoric;' they say nothing about the human condition which could not be said with the term 'organizational communication,' and they say nothing at all about the *meaning* of collective life, about 'progress' and 'human destiny' (1980b, p. 240).[3] This is also evident in Simon's 1991 commentary in which he praises Stewart's study of the Knights of Labor (Stewart, 1991), in which Stewart focuses on the group's internal rhetoric, as a study "most consonant with my own theoretical framework" (1991, p. 95). This study, then, is an exemplar of managerial rhetoric and the conventional sociological approach.

Simons' expansion of his definition combined with his conceptual reduction of social movements to organizations subverts his framework for the rhetorical study of social movements. By expanding his definition to include top-down social movements, Simons breaches the firewall he had established between bottom-up noninstitutionalized collectivities and institutionalized organizations in order to justify and maintain the distinctiveness of social movement studies. This firewall is key because Simons' own conceptualization of social movements renders them practically indistinguishable from institutionalized organizations. This is evident in his approving adoption of the sociological terms "social movement organization" (SMO) and "social movement industry" (SMI) (Simons et al., 1984, p. 793) and in his analysis of social movements through the prism of formal organizations: "A social movement is not a formal structure, but it nevertheless is obligated to fulfill parallel functions. Like the heads of private corporations or government agencies, the leaders of social movements must meet a number of rhetorical requirements" (1970, p. 3; Simons et al., 1984, p. 807).

Because he had reduced the study of the rhetoric of social movements to organizational communication in all but name, for over 20 years (1970–1990) Simons' claim to a distinctive domain of study rested on a two-step argument: (1) "that movements are in a uniquely precarious position due to their non-institutionalized, bottom-up status" (Simons et al., 1984, p. 807); and (2) this status uniquely distinguishes social movements from established institutions and is of crucial theoretical significance (Simons, 1980, pp. 308–309; Simons et al., 1984, pp. 841, 844). Indeed, it enabled Simons, in stark contrast to the many theorists who offer a plethora of top-down rhetorical theories, to offer a bottom-up rhetorical theory (Simons, 1980, p. 309).

Recognizing this distinction as the keystone of his rhetorical theory of social movements, Simons had been vigilant in protecting it. When Zarefsky had the temerity to suggest it was a distinction that did not make a difference (1977, 1980), Simons hammered Zarefsky, accusing him of ignoring his own work (Zarefsky's), unwittingly undermining his own argument, and threatening to significantly impede movement scholarship (Simons, 1980, pp. 306, 312–314; Simons et al., 1984, pp. 844–845).

Given the singular importance of the bottom-up distinction in Simon's framework for social movement study, his admission of top-down institutionalized social movements signals the collapse of his rhetorical theory of social movements into organizational communication, clearing the field for the construction of a *rhetorical* theory of social movements.[4]

MAKING THE INVISIBLE MATTER

My critique of Simons' work is not limited to the charge of theoretical inconsistency because theories have consequences. I am vitally concerned about the practical consequences of the hegemonic model of social movement study on the discipline of rhetorical studies. If theories are frames for seeing and making sense of the world, what does Simons' framework enable us to see and, more important in our case, what does it obscure? What stuff, what activities and groups, can rhetorical theory not make sense of from Simons' perspective? And just as invisible matter is the key to contemporary physics, could what has been invisible or marginalized within the dominant paradigm offer clues about how to reconstruct a rhetorical theory of social movements?[5]

With Simons' framework as a guide, we would focus on collectivities that resemble institutionalized organizations, paying particular attention to their resources and the rhetoric their leaders employ to fulfill functional requirements with the greatest organizational efficiency. The rhetoric, then, follows from the demands of the organizational form and operates within organizational matrices.

This is an dangerously reactionary move because many groups working toward social change adopt alternative forms both for reasons of principle and as part of their social critique. It is also ironic for Simons to proffer an insider form of social organization as the model for studying social movements, considering that in a 1972 essay he argued so eloquently for coercion as a legitimate form of persuasion and condemned so stridently the establishment bias in the traditional rhetoric for insiders as being inappropriate for the study of social conflict and groups agitating for social change.

Frameworks are also blinders, and what we miss with respect to social movements by using Simons' framework is almost everything of importance. Specifically, we cannot make sense of the radical environmental groups and their rhetorical tactics, which are a driving force behind environmentalism and are challenging and changing modern industrial society's consciousness with regards to nature and humanity-nature relations, because, as organizations, radical environmental groups are relatively insignificant, if not nonexistent. Even Greenpeace, whose size and wealth may merit it consideration as an organization, yields little to analysis from an organizational perspective. This is true in part because many of Greenpeace's most important acts and victories (against nuclear testing, whaling, seal hunting) were performed when Greenpeace was, at best, a ragtag group. Even today, Greenpeace functions as an organization that actively works to subvert itself as an organization. As Greenpeace USA media director Peter Dykstra notes, "The people who are drawn to this organization hate structure" (quoted in Ostertag, 1991, p. 86). Indeed, the "purpose of the organization is to be campaign-driven" (Greenpeace USA director Peter Bahouth, quoted in Ostertag, 1991, p. 87). As Greenpeace International director David Mc-Taggart makes clear, "Look, we were never made to be a bureaucracy. We're made to be expendable, and that is a fucking fact. I hope we're big enough and strong enough now to take on some issues big enough to maybe shut us down, or at least a part of us" (quoted in Horton, 1991, p. 112).

Not driven by the organizational imperative of self-preservation, neither is Greenpeace concerned with the establishment of a "hierarchy of authority and division of labor . . . in which members are persuaded to take orders" (Simons, 1970, p. 3). Instead, "most Greenpeace projects are generated at the bottom of the organizational pyramid, by campaigners, who can refuse to take part in activities of which they do not approve. 'All campaigning must be individual acts of conscience'" (Harwood, 1988, p. 76). As organizational researcher Ronald Shaiko observes, Greenpeace is a "hyperdecentralized organization" driven by issue campaigns, with "planning and strategy . . . based on demands and needs as they arise," and "little task-based division of labor" (1993, p. 96). According to Shaiko, "Clearly, if one were to impose a standard of organizational effectiveness, utilized when evaluating mainstream environmental organizations in the United States, Greenpeace U.S.A. would fail miserably; it would be looked upon as a $25 million money pit" (1993, p. 96). Greenpeace's violations of the conventional model of organizational form and effectiveness is not a failure but a message about and a challenge to the dominant norms of the social system.

If Greenpeace resists organizational analysis, Earth First! and environmental justice groups are openly hostile to such a perspective, being small groups that are only sometimes linked through networks and that are passionately anti-organizational. For example, Earth First! is an anti-hierarchical organization with no official leaders, no national headquar-

The crew of the first Greenpeace mission, intended to protest nuclear weapons testing by the United States on Amchitka in the Aleutian Islands. In attempting to "bear witness" to the explosions, the ragtag group brought political pressure that convinced the United States to discontinue testing there.

ters, no membership lists, no dues, no board of directors, and no tax-exempt status (Setterberg, 1995, p. 70; Kane, 1987, p. 100). This was a conscious decision, as Earth First! cofounder Dave Foreman explains: "Earth First! decided to be non-organizational: no officers, no bylaws or constitution, no incorporation, no tax status, just a collection of women and men committed to the Earth. . . . We felt that if we took on the organization of the industrial state, we would soon accept their anthropocentric paradigm, much as Audubon and the Sierra Club already had" (1991a, p. 21).

In describing women leaders in the environmental justice movement as "carrying out an invisible revolution" (1997, p. 1), Temma Kaplan ascribes their invisibility to the media bias for the spectacular and the academic bias whereby "sociologists and political scientists who analyze collective action generally focus on formal organizations and underrate the significance of the kinds of activities in which the women considered here engage . . . informal ties such as friendship networks and connections among church members" (1997, p. 181).

By adhering to Simons' sociological framework for studying social movements, wherein social movements are reduced to organizations,

heorists leave themselves incapable of making sense of one
important social movements of our time, the change in pub-
ness in modern industrial societies of the meanings of nature, progress, reason, and even humanity, changes propelled by the
rhetorical acts of small groups or even lone individuals amplified by
modern means of mass communication so that they can reach millions
of people.

OUTLINES OF A RHETORICAL THEORY OF SOCIAL MOVEMENTS

Although Simons' approach leads us away from rhetoric, the most important developments in sociology for the study of social movements—
the identity-oriented paradigm—point to the necessity of a rhetorical
perspective for making sense of new social movements. This move in sociology is consonant with the more general discursive turn in the social
and natural sciences, which, in many ways, is a recognition that social
life rests on constructed, contingent rhetorical foundations (Nelson,
Megill, & McCloskey, 1987; Simons 1989). If we understand rhetorical
activity as constitutive of the social and political collectivity, a rhetorical
theory of social movements, as opposed to Griffin's historical theory or
Simons' sociological theory, is necessary to explain social change. From
such a rhetorical perspective,

> A movement does not "move" in the objective world. It can only be
> *interpreted* through bits and pieces of behavior and "created" by the
> symbolic form and meaning these verbal and non-verbal behaviors
> take on in relationship to already established symbolic forms and
> meanings. A movement is perceived, created, and responded to *sym-*
> *bolically* as its confrontational strategies are juxtaposed with the sym-
> bolic forms and contents of the established and legitimized collectives
> with which it interacts. (Cathcart, 1980, p. 268)

Tellingly, what McGee and Cathcart argue must be the focus of a
rhetoric of social movements—changes in human or collective consciousness or changes in the symbolic interpretation of the environment—is precisely what Simons chooses to exclude. Simons cites
Blumer's discussion of general social movements as cultural drifts, "a
general shifting in the ideas of people, particularly along the lines of the
ideas people have of themselves, and of their rights and privileges" only
to "arbitrarily exclude Blumer's 'cultural drifts' from our formal definition of social movements" (Simons et al., 1984, p. 793). Only Cathcart,

with his theorizing of confrontational rhetoric, and McGee, with his focus on social movements as meanings, attempt to present a rhetorical theory of social movements that focusses on and can make sense of changes in the collective consciousness of a culture, changes in the symbolic interpretation of the environment or grid of intelligibility. Thus, it is to their work that we must now turn.

In the shadows of Simons' sociological edifice (and in response to it), Cathcart laid the groundwork for a rhetorical theory of social movements. Cathcart's position emerges out of his noninstrumental, constitutive understanding of rhetoric as "a way of being and a way of knowing . . . our only means of constructing social reality and maintaining social control" (1983, pp. 69, 70). In a series of essays (1972, 1978, 1980, 1983) Cathcart's position evolved from the notion that "movements are essentially rhetorical" (1972, p. 86) through the contention that "movements are a kind of ritual conflict whose most distinguishing form is *confrontation*" (1978, p. 235), with confrontational rhetoric, in contradistinction to managerial rhetoric, being rhetoric that "challenges the system's values and its perfectability" (1980, p. 268), to the strong claim that social movements are not objective phenomenon but symbolic transformations of reality effected by rhetorical forms and languaging strategies (1983, p. 70). Social movements, then, are changes in social reality through changes in meanings of "foundational" rhetorical forms, such as ideographs. Consequently, from Cathcart's perspective, rhetoricians should focus "movement studies on the languaging strategies—call them motives, symbols, or ideographs" (1983, p. 73) through which people construct, maintain, and change social reality.

Although Cathcart has made a promising start, his thinking is dogged by the sociological conception of social movements as empirical things, groups or organizations, so that his work retains the dual and dueling conceptions of social movements. On the same page on which Cathcart argues that movements do not move in the objective world and, indeed, are created only through language, he also asserts, "Movements do exist in space and over time. They have membership, leadership, and organization" (1980, p. 268). In a later essay, after explicitly arguing against studying social movements as objective phenomena (1983, p. 70), Cathcart insists on conceptualizing social movements as collectives (1983, pp. 71–74). Cathcart's inability to shed the notion of social movements as collectivities hampers the development of his rhetorical theorizing of social movements, so that in too many respects social movements for Cathcart are groups the public finds threatening to the system (1980, p. 270). Indeed, the contradiction in Cathcart's work is so glaring that McGee, his ally in the construction of a rhetorical theory of social movements, chalks it up to "a slip of Cathcart's typewriter" (1983,

p. 74). The point is not that groups do not exist, just that they are not the social movements themselves. Instead, groups, as well as individuals or institutions, through their rhetorical tactics and strategies create social movements, changes in public consciousness with regards to a key issue or issues, measurable through change's in the meanings of a cultures key terms in public discourse.

Starting in 1975, McGee (1975, 1980c, 1983) offers the first rigorously rhetorical theory of social movements. McGee begins by burning the bridges others had constructed to sociology and history, declaring that by adopting the empirical sociological definition of social movements, "sociologists, social psychologists, and most contemporary rhetoricians seem to be studying 'movement' only by a stretch of the imagination" (1975, p. 235). For McGee, social movements are not phenomena but sets of meanings. Consequently, a rhetorical theory of social movements is not a study of organizational communication but an account of the social consciousness of a society. Changes in the social consciousness are empirically present in the public discourse or rhetoric used to describe "reality." In short, social movements are changes in the meanings of the world, redefinitions of reality, with such realities always being constructed through the filter of rhetoric (McGee, 1975, p. 243). Although Simons found the tools of rhetoric insufficient for studying cultural drifts, McGee and Cathcart turn to the relatively new tool of ideographs. They argue that cultural drifts or social movements in human or collective consciousness can be proven by "observing changes in the 'ideographic' structures of social norm-systems" (McGee, 1980c, p. 243; see also Cathcart, 1983, pp. 72–73). In other words, social movements are materially manifest not in groups but in public discourse such as ideographs.

Recall the definition in McGee's seminal essay:

> An ideograph is an ordinary-language term found in political discourse. It is a high-order abstraction representing collective commitment to a particular but equivocal and ill-defined normative goal. It warrants the use of power, excuses behavior and belief which might otherwise be perceived as eccentric or antisocial, and guides behavior and belief into channels easily recognizable by a community as acceptable and laudable. (1980a, p. 15; see also Condit and Lucaites, 1993, pp. xii–xiii)

Although the diachronic dimension of ideographs is important in setting parameters and understanding communal resources for discursive interventions, McGee argues that the synchronic dimension is paramount in that it enables us to understand how ideographs, "meant to be taken together, as a working unit" (McGee, 1980a, p. 13), function *presently* as rhetorical forces. Even though both dimensions are important, a diachronic analysis yields merely a formal grammar (McGee, 1980a, p. 12) devoid of the force and currency of a synchronic analysis (McGee, 1980a, p. 14), for "time is an irrelevant matter in *practice*" (McGee, 1980a, p. 12).[6]

Despite McGee's privileging of the synchronic structure of ideograph clusters, most important analyses of ideographs focus on the diachronic dimension or the synchronic dimension of a particular ideograph in isolation (McGee, 1980b; Condit and Lucaites, 1993; Charland and Lucaites, 1989). For example, Condit and Lucaites devote a book to the diachronic structure of equality. They pay some attention to the synchronic cluster of equality, liberty, and property in the 1700s and in their Afterword they touch upon a possible "synchronic locus of Equality in American public discourse in the years ahead" (1993, p. 220), but their clear goal (admirably achieved) is to offer "a *rhetorical history* of the American ideograph Equality" (1993, p. xvii; see also pp. xviii, 217).

I want to suggest that this neglect of the synchronic structure, of the linking of ideographs, is not accidental but rather points to a certain lacuna in McGee's theorizing of ideographs. Although McGee tentatively employs the vague notions of connector, circumstance, contradiction, dynamic consonance (1980a, pp. 13–14), and myth (1975, pp. 246–247) to attempt to explain how the synchronic structures of ideographs come into being and persist, his work needs to be supplemented by Laclau and Mouffe's subsequent theorizing of articulation, antagonism, and hegemony in order to explain how ideographs, in a social field marked by contingency instead of necessity, are linked in certain synchronic structures and what conditions enable groups to challenge or even change a particular linking of ideographs, that is, the hegemonic discourse.

SUPPLEMENTING IDEOGRAPHIC ANALYSIS

Laclau and Mouffe "begin by renouncing the conception of 'society' as founding totality" (Laclau and Mouffe, 1985, p. 95). They argue for the openness of the social and the impossibility of fixing ultimate meanings. Discursive structures constitute and organize social relations and are the result of articulatory practices (Laclau and Mouffe, 1985, p. 96). Articulation is

> any practice establishing a relation among elements such that their identity is modified. . . . The practice of articulation, therefore, consists in the construction of nodal points which partially fix meaning

and the partial character of this fixation proceeds from the openness of the social, a result, in its turn, of the constant overflowing of every discourse by the infinitude of the field of discursivity. (Laclau and Mouffe, 1985, p. 105, 113)

Articulation has two aspects: speaking forth elements and linking elements. Though elements preexist articulations as floating signifiers, the act of linking in a particular articulation modifies their character such that they can be understood as being spoken anew. The linking of elements into a temporary unity is not necessary, but rather is contingent and particular and is the result of a political and historical struggle (Angus, 1992, pp. 540–543, 552–553).

The growing awareness of the struggle over meaning and the openness of the social can be traced to the rise of instrumental reason (Horkheimer, 1947, 1972; Horkheimer and Adorno, 1972). Although traditional organizing principles established necessary relations among elements or moments in a closed totality, instrumental reason revealed traditional organizing principles, including religions as well as metaphysics, to be arbitrary, one among many in a field of competing organizing principles. Instrumental reason claims to be a universal logical totality, but its claim is undermined by its inability to theorize ultimate ends and by its presupposition and obscuring of the immediate, experienced lifeworld (Angus, 1984). Absent a universal organizing principle or grand narrative, what is left is the field of discursivity, wherein numberless discourses compete to articulate free-floating elements into hegemonic unities. To lend some concreteness to this discussion and to help clarify Laclau and Mouffe's theory of articulation, let us trace the diachronic structure of the key element "domination of nature." For our purposes, this is a key element due to its centrality in the discourse of industrialism and environmental politics.

Leiss, in his book *The Domination of Nature* (1974), traces the history of that element. His book is particularly useful because he perceives the idea of the domination of nature as part of a dynamic historical process. Neither the element—the domination of nature—nor the articulations into which it has entered have remained static. The history of the domination of nature is traced by Leiss, and others, notably Lynn White Jr., in "The Historical Roots of Our Ecological Crisis" (1968), to the Judeo-Christian tradition and the book of Genesis: "Be fruitful and multiply, and replenish the earth and subdue it: and have dominion over the fish of the sea and over the fowl of the air and over every living thing that moveth upon the earth" (Bible, Gen. 1:28 King James Version). Such a perspective supports a conception of land as a commodity both distinct from and belonging to humans (Leopold, 1949/1968, pp.

viii, 204–205). Nonproductive land, wilderness, was conceived as cursed and desolate, an arid wasteland, "the abode of demons and devils . . . the earthly realm of powers of evil" (Nash, 1967/1973, pp. 15, 17). The Catholic Church branded as heretical Saint Francis of Assisi's biocentric teachings that granted wild creatures souls (Nash, 1967/1973, p. 19). As Leiss notes, however, initially the domination of nature was constrained by the religious discourse in which it arose (this is similar to Horkheimer's "objective reason"): "Man's will is not the highest principle in heaven or on earth, but instead is checked and limited by ethical norms established independently of it. Similarly, the surrounding world of nature has a purpose entirely apart from its function as the basis of human activity: it is a divine creation and therefore sacred" (1974, p. 34). Further, the Christian tradition also conceptualizes wilderness as a sanctuary, "a place of refuge and religious purity . . . for freedom and the purification of faith" (Nash, 1967/1973, pp. 18, 16). Domination of nature was constrained in a discourse involving God, humanity, humanity's place in the hierarchy of beings, and nature as both material provider and divine creation. The breakdown of this articulation allows the extreme realization of the domination of nature in the discourse of industrialism.

Leiss cites the work of Francis Bacon as the point at which the religious discourse constraining the domination of nature started to crumble. Bacon was working within a religious context but was able to carve out (to disarticulate) a niche for science: "For man by the fall fell at the same time from his state of innocency and from his dominion over creation. Both of these losses however can even in this life be in some part repaired, the former by religion and faith, the latter by arts and sciences" (quoted in Leiss, 1974, p. 49). Dominion over creation through science promised material wealth. Bacon not only legitimated science in a religious context, he separated scientific knowledge and moral knowledge, facts and values, the scientific world and the everyday life-world, instrumental reason and objective reason.

> For Bacon and his contemporaries religion had provided the means for understanding science as a human activity . . . religion supplied the link uniting scientific activity with everyday action in the life-world. . . . But the overwhelming success of the marriage between industry and the new science, and the growing social authority of the novel scientific methodology, spelled inevitable defeat for the traditional scheme of religiously based ethics. (Leiss, 1974, p. 134)

The rise of instrumental reason and the consequent waning of religion unhinged the domination of nature from its link to Christianity.

Nature became separated into two spheres: "intuited nature . . . the experienced nature of everyday life" and "scientific nature . . . abstract-universal, mathematized nature" (Leiss, 1974, pp. 135–136). As modern science, characterized by instrumental reason, grew in importance, its mathematized view of nature as a "silent, colorless universe of matter in motion" (Leiss, 1974, p. 132) became the definition that counted. With this focus on a mathematical nature, other aspects of nature—sacred nature, nature with intrinsic worth, and nature as the background of everyday life—faded in importance. Subsequently, dominion over a universe of matter in motion, mere stuff, entailed different possibilities of action than dominion over a sacred, divine creation.

The domination of nature was then linked to instrumental reason and technology, not God, and rearticulated under the discourse of industrialism—humanity, by dominating nature through the use of instrumental reason and technology, will achieve progress. This example of the domination of nature, then, illustrates the key point that an element such as the domination of nature is both thematized from the lifeworld and constituted anew in an articulation. An element is not a fixed identity and does not have an essential meaning.

The practice of articulation can be understood as both attempting to fix meaning within the field of discursivity and as attempting to fix the context—"an attempt to dominate the field of discursivity, to arrest the flow of differences, to construct a centre" (Laclau and Mouffe, 1985, p. 112). For instance, industrialism could be considered a hegemonic discourse that temporarily defines the field of discursivity. Marxism and capitalism are two competing discourses. To simplify, they are fighting over who should own the factory. Neither questions whether the factory should be built in the first place (nor whether nature should be conceived as a storehouse of resources). In a fundamental sense, they both operate within the taken-for-granted context of industrialism.[7]

Antagonisms

Antagonisms make possible the questioning, disarticulating, and rearticulating of a hegemonic discourse. Antagonisms point to the limit of a discourse. An antagonism occurs at the point of the relation of the discourse to the surrounding lifeworld and shows the impossibility of the discourse constituting a permanently closed or sutured totality. It shows the linkage of elements to be contingent, not necessary. "Antagonism as the negation of a given order is, quite simply, the limit of that order" (Laclau and Mouffe, 1985, p. 126). For example, during the history of the United States, the "American Dream" (land of freedom and opportunity for all) has faced antagonisms (slavery, segregation, oppression of women, exploitation of workers) that have exposed the limits of the "American Dream" and led to struggles (the abolition movement, the Civil War, the woman suffrage and women's liberation movements, the labor movement, the civil rights movement) that disarticulated and rearticulated the "American Dream."

To take an example with respect to industrialism, the threat of nuclear holocaust forces us to rethink the linkage of instrumental reason, technology, the domination of nature, and security. The ultimate product of industrialism, nuclear weapons, built in order to ensure security, makes possible the opposite—extinction—and so radically subverts the promised progress of industrialism. This antagonism gave rise to nuclear disarmament groups that tried to rearticulate the notion of security. Today, antagonisms such as global warming, ozone depletion, toxic waste sites, mass extinctions, pesticides in food and water, and so on also provide the opportunity to question and disarticulate industrialism.

It is important to make clear that there are no original or essential antagonisms, but that antagonisms emerge as limits from within the social: "The limit of the social must be given within the social itself as something subverting it, destroying its ambition to constitute a full presence. Society never manages fully to be society, because everything in it is penetrated by its limits, which prevent it from constituting itself as an objective reality" (Laclau and Mouffe, 1985, p. 127).

Laclau and Mouffe take feminism as an example. To point to the biological difference between women and men as an original or foundational antagonism makes the problem of sexism unanswerable. Rearticulation becomes impossible. Linkage with other social struggles is not possible. If feminism sees men as the problem and the civil rights movement sees white people as the problem, there would be no common ground that would enable the two struggles to unite and understand how their particular oppressions expose a common relation of dominance within the hegemonic discourse under which both are oppressed. Antagonisms are specific, not foundational. They are the recognition of differences as socially constructed—as the result of the practice of articulation. Indeed, a foundational antagonism is not an antagonism as such but just a difference. It has always been that way, and the situation will remain the same after the perception of difference. The difference of sex or skin color cannot be changed. The antagonism due to relations of domination constructed around sex or skin color is the arena for struggle: "The political space of the feminist struggle is constituted within the ensemble of practices and discourses which create the different forms of the subordination of women; the space of the anti-racist struggle, within the overdetermined ensemble of practices constituting racial discrimination" (Laclau and Mouffe, 1985,

p. 132). With respect to humanity/nature, humanity's domination of nature is not natural but rather is constructed through articulatory practices.

Antagonisms make evident the limits of a discourse. However, antagonisms are differences, limits, in a hegemonic discourse that must be articulated by groups as antagonisms in order to subvert the hegemonic discourse. Antagonisms occur when previously construed "natural" relations of subordination are articulated as socially constructed relations of oppression and domination. For example, within the discourse of industrialism, toxic waste dumps, strip-mined land, and other environmental problems are "the price of progress," the normalized cost of economic growth, and the people affected need to sacrifice for the common good. The alternative discourse of environmental justice, however, subverts industrialism by enabling the mother whose son is sickened by going to school on top of a toxic landfill (Lois Gibbs, 1982, 1993), the teenager of color who sees that it is always her people's community that is targeted for dumps and incinerators (Rhiannon Chavis-Legerton, 1993), and the rural family whose land is threatened by a distant corporation's strip-mining operation (KFTC members) to understand their relations of subordination to the hegemonic discourse of industrialism as relations of oppression that expose the limits of industrialism. The price they are being forced to pay is rearticulated as class discrimination, institutional racism, and corporate colonialism.[8]

There is a slippage in Laclau and Mouffe's usage so that the linkage of various antagonisms in a chain of equivalences is also called an antagonism. For a black person in the 1950s, not being allowed to work, live, travel, ride, eat, or drink in the same places as whites came to be seen as equivalent signs of a general oppression, sparking the civil rights movement. The threat of global warming, the local toxic waste site, pesticides in one's food, and the specter of a nuclear holocaust may be seen as relatively equivalent, a chain of equivalences, insofar as they point to the hegemonic discourse of industrialism and spark a struggle to disarticulate its linkages.

On another level, the antagonisms of various social struggles retain their particularity but are also linked as equivalent in that they all point to the limit of the dominant hegemonic discourse. The different antagonisms that give rise to workers' struggles, feminist struggles, anti-racist struggles, and so on all make possible the disarticulation of the hegemonic articulation that constructs these various groups in relations of oppression. To enable distinction, I will term a chain of equivalences an "antagonism chain." For Laclau and Mouffe, the task of the new social movements involves "expanding the chains of equivalents between the different struggles against oppression" (1985, p.176).

SUPPLEMENTING ARTICULATION THEORY

Elements, which are linked by articulation, remain untheorized by Laclau and Mouffe and, instead, are simply defined as "floating signifiers, incapable of being wholly articulated to a discursive chain" (1985, p. 113). In a later essay Laclau (1993a) describes floating signifiers as relatively empty and ambiguous. Further, he argues that the "ambiguity of the signifier 'democracy' is a direct consequence of its discursive centrality; only those signifiers around which important social practices take place are subject to this systematic effect of ambiguity" (1993a, p. 342). Other examples of such elements include "unity," "revolution," and "woman." The meaning of an element is always contingent and relational: "The signifier 'woman' in itself has no meaning. Consequently, its meaning in society is going to be given only by a hegemonic articulation" (Laclau, 1988, p. 255), and its relation to other elements in a discourse will modify its meaning.

Laclau and Mouffe's description of elements bears more than a passing resemblance to McGee's concept of ideographs. McGee, however, offers a more complete theorization of ideographs, so his term should be incorporated into a theory of articulation.[9] The significance of this move is not merely to gain a better sense of the term "elements" but to supplement a lacuna in Laclau and Mouffe's work: the absence of the role of rhetoric. Rhetoric is absent in explaining both the construction of hegemonic discourses (i.e., how one articulates a hegemonic discourse in a postmodern social field not ruled by a logic of a priori essentialism) and how discursive elements compel beliefs and actions (i.e., why elements and the linking of elements have social consequences). McGee's concept of ideographs gives elements rhetorical force.

Similar to Laclau and Mouffe in their description of elements, McGee conceptualizes ideographs as terms crucial to important social practices: "The ideology of a community is established by the usage of such terms in specifically rhetorical discourse. They are the basic structural elements, the building blocks, of ideology" (McGee, 1980a, pp. 16, 7), whose meanings do not rigidify, but rather are understood relationally. As McGee puts it, "An ideograph, however, is always understood in its relation to another" (1980a, p. 14). McGee adds to Laclau and Mouffe's work by insisting that ideographs are not philosophical abstractions but words that exist in real discourse (1980a, pp. 9, 7) and "that function as guides, warrants, reasons, or excuses for behavior and belief" (1980a, p. 6). As rhetorical forces, ideographs are "definitive of the society we have inherited, they are *conditions* of the society into which each of us is born, material ideas which we must accept to 'belong'" (McGee, 1980a, p. 9).

McGee's discussion of the diachronic structure of ideographs adds specificity to Laclau's discussion of elements as floating signifiers with a "degree of emptiness" (1993a, p. 342). Although McGee agrees with Laclau and Mouffe that the meanings of ideographs are ambiguous and relational, the dimension of diachrony suggests a degree of fullness. McGee talks of emptying and filling ideographs in political struggle, so that the meaning of an ideograph is articulated in a synchronic cluster or hegemonic discourse. The history of usages of an ideograph, however, establishes the parameters for its meaning within a current discourse (1980a, p. 16).

McGee's contention that the "awareness of the way an ideograph can be meaningful *now* is controlled in large part by what it meant *then*" (1980a, p. 11) meshes well with Laclau's claim that political argument in a social field permeated by contingency, that is, society not as ground but as argumentative texture (1993a, p. 341), must be characterized by phronesis. Although a turn to phronesis can be construed as conservative (Schlechtweg, 1990, p. 22) in the reflexive way that any valuing of tradition can be considered conservative, it is Aristotle's conservative definitions of virtue, right action, and tradition that render it so (Aristotle, 1962/1981, pp. 3–18, 38–44, 152–173; Warnick, 1989, pp. 305–307). There is a tradition of exercise of power, but there is also a tradition of struggle (Laclau, 1993a, p. 341). For example, environmental activists are part of a tradition of radical direct action that can be traced back at least as far as the Luddites (Thompson, 1964; Sale, 1995), if not to the Diggers (Hill, 1967, 1972, 1985). The characteristics of phronesis that interest Laclau are its open-endedness, pragmatism, emphasis on the particular, embeddedness in the local and historically specific, and recognition of contingency (Laclau, 1993a, p. 341; Warnick, 1989, p. 306; Charland, 1991, p. 73). This is in contrast with the instrumental reason of science, which emphasizes the universal, abstract, and timeless. Phronesis, then, is the "intelligent understanding of contingency" (Charland, 1991, p. 72) that guides praxis in an open social field.

Their mutual turns to phronesis can serve as a bridge linking Laclau and Mouffe's poststructuralist discursive theory of politics with McGee's rhetorical theory of social change. As the previous pages suggest, it is a fruitful meeting. Laclau and Mouffe's theorizing of articulation enable us to start thinking how synchronic structures of ideographs are constructed, maintained, and transformed. Likewise, McGee's theorizing of ideographs fleshes out Laclau and Mouffe's vague notion of elements, imbuing their philosophical speculations with rhetorical force and enmeshing their abstract work in daily political struggles over the rhetorical terms that define our worlds.

CHAPTER 3

IMAGING SOCIAL MOVEMENTS

Given this retooled postmodern rhetorical theory of social movement, this mutually supplementing amalgamation of McGee's work with Laclau and Mouffe's, what do we do? Following his claim that social movement is materially manifest in the public rhetoric of a society, particularly in such forms as ideographs, McGee suggests studying the rhetorical tactics of groups that have changed or that may change the meanings of key ideographs, thus changing human consciousness. Following the previous rereading of McGee's work in conjunction with Laclau and Mouffe's work, I want to suggest we can and should study the rhetorical tactics of groups attempting not merely to move the meanings of key ideographs but to disarticulate and rearticulate the links between ideographs, the synchronic cluster or discourse. Radical environmental and environmental justice groups meet this criterion. My next step is to decide what linkage of ideographs, what discourse, are they attempting to move the meaning of, to rearticulate. In my judgment, these groups aim to challenge and transform the hegemonic discourse of modern society, the previously mentioned discourse of industrialism: Humanity (universalized Western rational "man"), by dominating nature (storehouse of resources, mechanistic object) through the use of reason (instrumental reason, science) and technology (autonomous and inevitable), will achieve progress (security, autonomy from nature, overcoming scarcity, ever increasing standard of living). In particular, they are concerned with the ideographs progress and nature and their linkages, though the ideographs humanity, technology, and reason and their linkages are also being challenged. Before exploring how the image events of radical environmental groups work to challenge the ideographs progress and nature, and thus attempt to disarticulate the hegemonic discourse of industrialism, I need to justify my choice of ideographs and discourse as the central rhetorical terms of modernism.

PROGRESS, NATURE, AND INDUSTRIALISM

The importance of the ideographs progress and nature is almost self-evident. Though it is beyond the scope of the present study to perform a diachronic analysis of any particular ideograph, and as I argued earlier the synchronic dimension is of more pressing significance, it is worth taking a brief historical look at this topic. After sketching progress and nature as ideographs, I will then cite some public usages of these ideographs in order to illustrate how ideographs as "forces" warrant certain beliefs and behaviors with regard to the world's environmental crisis before focusing on image events as both attempts to disengage the contemporary articulation of progress and nature and to "change the structure of ideographs and hence the 'present' ideology" (McGee, 1980a, p. 13).

Belief in progress is contemporary common sense. Progress has become the taken-for-granted background of Western culture through a historical and political process. "No single idea has been more important than, perhaps as important as, the idea of progress in Western civilization for nearly 3,000 years" (Nisbet, 1980, p. 4). By extension, as Western culture has spread over the world, so has the ideograph of progress. For most of those three thousand years, progress was an ideograph among other ideographs. During the Industrial Revolution, however, the ideograph progress achieved dominance and became "the 'title' or 'god-term' of all ideographs, the center-sun about which every ideograph orbits" (McGee, 1980a, p. 13). As Nisbet notes, "The concept of progress is distinct and pivotal in that it becomes the developmental *context* for these other ideas" (1980, p. 171). R. G. Collingwood relates how the 19th century saw the institutionalization of the concept of "progress as guaranteed by a law of nature" (1956, p. 322).[1]

Ideographs are dynamic, that is, their meanings or sum of an orientation change over time in response to the pressures of rhetoric and circumstances (McGee, 1980a, p. 13). Progress once meant spiritual, moral, intellectual, and scientific progress. These meanings have been eclipsed, especially since the horrors of the World Wars, during which instrumental reason made genocide more efficient and science made nuclear holocaust possible. Heisenberg's Uncertainty Principle and Godel's Theorem also undermined faith in scientific progress by suggesting the limits of knowledge. In a few decades, then, the advances of technology and theoretical physics—both privileged markers of "civilization"—led to senseless slaughters on an unprecedented scale and the possibility of the nuclear obliteration of humanity. The ideograph progress, however, has remained a potent force warranting behavior and belief, but now with a primarily economic meaning. Indeed, Christopher Lasch, in his history of progress, which he deems the working faith of our civilization, argues that it is only the defining of progress as economic growth that enables, in the 20th century, "the persistence of a belief in progress in a century full of calamities" (1991, pp. 47, 45, 52, 13). Some critics, such as the Agrarian Lyle Lanier, even argue that the "gospel" of progress has itself become an economic commodity: "Modern industrialism has found the use of 'progress,' as a super slogan, very efficacious as a public anaesthetic . . . perhaps the most widely advertised commodity offered for general consumption in our high-powered century" (1972, p. 123).

Raymond Williams opens up a discussion of the ideas of nature with the assertion that "the idea of nature contains, though often unnoticed, an extraordinary amount of human history. . . . The idea of nature is the idea of man . . . of man in society, indeed the ideas of kinds of societies" (1980, p. 71). He then goes on to list some of the ideas of nature: goddess, mother, monarch, machine, red in tooth and claw, and so on. The key point here is not merely that nature is a social category whose meaning is culturally defined, but rather that the various meanings of the ideograph nature do ideological work, buttressing certain beliefs, warranting actions, justifying forms of society, and naturalizing hierarchical social relations. As Mary Douglas explains, nature functions as a coherent principle of social control in support of a particular social order (1975, pp. 239, 243).[2] For example, visitors to the Akeley African Hall of the American Museum of Natural History in New York City are confronted by the "Giant of Karisimbi," a silverback gorilla in an aggressive pose protecting his "family": a mother, baby, and juvenile. Mounted in the 1920s, this diorama enlists nature and primatology in support of the patriarchal nuclear family. (For a powerful reading of the Akeley African Hall, see Haraway, 1989, pp. 26–58.)

In Western culture, the "fathers" of modern science, Bacon, Descartes, and Galileo, constructed nature as an object to humanity's subject, a machine, matter in motion. This disenchanted, scientific nature has held sway as the dominant meaning of nature ever since and has been crucial for the projects of science and industrialism.[3] In short, as Neil Evernden argues in *The Social Creation of Nature*, "The first step in establishing a new social order appears to be the construction of a new nature that will justify, even demand, its implementation. . . . Nature is reformulated to become the kind of entity that will demonstrate the norms we wish to discover, and both nature and biology are pressed into the service of social reality" (1992, p. 16). In the present cluster of ideographs, nature is a storehouse of resources used to fuel the engine of industrialism.

Progress and nature, along with the other ideographs in the discourse of industrialism, define our society for us, justify certain beliefs and actions, and signify collective commitments, such as the belief in the necessity and possibility of unlimited growth, the belief in technology as the answer to all problems (including spiritual and environmental problems), and the treatment of all nonhuman life forms as resources to be exploited (certain groups of humans get defined as nonhuman in certain circumstances). These ideographs are present in many different forms of discourse, as even a cursory skimming of the cultural surface reveals. President Bill Clinton, in his address to the United Nations on September 27, 1993, says "We are common shareholders in the progress of humankind" (quoted by Cockburn, 1993, p. 486). Former president Ronald Reagan in an earlier career was host for *GE Theater,* and as we all know, "At General Electric, Progress is our most important product" (Chasnoff, 1991). Clark Kerr, in *Industrialism and Industrial Man,* describes the "progress of industrialism as an overwhelming historical juggernaut that refashions the whole world to suit its purposes" (quoted in Winner, 1986, p. 167). In the ongoing debate about who will control the "information highway," MCI Communications chief Bert Roberts invokes progress to argue that "government should do everything possible to stay...far away from making any decisions. We have seen too many times when government involvement will thwart progress" (quoted in Burgess, 1993, p. D9). Robert Berzak of Union Carbide summons progress to defend his corporation against criticism of its actions concerning the Bhopal catastrophe, arguing that Union Carbide is dedicated to "industrial progress." William Safire, in a column praising the political potential of the information highway, opens by humming a hymn to progress: "Zipping along at 186 m.p.h. from Paris to London, traveling between those capitals in the same three hours as the Metroliner takes to go from New York to Washington, even a neo-Luddite has to salute the progress of technology" (1996, p. A15).

Even those who decry the consequences of progress bow to its inevitability. Aldo Leopold, a prophet for a new order of human–nature relations based on the "land ethic" (1949/1968) and a trenchant critic of the "high priests of progress" (1949/1968, p. 100) and the creed of *"salvation by machinery"* (quoted in Oelschlaeger, 1991, p. 222), still acknowledged in the opening to his canonical environmental text, *A Sand County Almanac*: "These wild things, I admit, had little human value until mechanization assured us of a good breakfast, and until science disclosed the drama of where they come from and how they live. The whole conflict thus boils down to a question of degree. We of the minority see a law of diminishing returns in progress; our opponents do not" (1949/1968, p. vii). A woman in Montana lamenting the imminent loss of the local forests to clear-cutting says, "I want something left out there for my grandkids. . . . We all would like to see things stay somewhat the same yet we know progress is coming" (National Public Radio, 1993). A rural community activist laments, "Western progress has marched across the millennia, dragging in its wake a hinterland increasingly global. White man's prayer shawls litter the Himalayan trails and his cola bottles decorate the dusty courtyards of African villages. From China to Chinatown, it's 'a burger, fries, and a forest to go' " (Bazan, 1996, p. 6A). *Washington Post* columnist Michael Wilbon mourns the passing of real basketball, symbolized by the demolition of Boston Garden, in favor of a brand of basketball characterized by advertising and profit, wherein then Orlando Magic star Shaquille O'Neal can describe his playoff matchup with Michael Jordan in the following terms: "I'll enjoy being on the same court with him. He gets publicity, I get publicity. There's enough marketing to go around for everybody" (quoted in Wilbon, 1995, pp. D1, D4). With resignation, Wilbon concludes, "It's almost as if the old Boston Garden simply gave way under the weight of all the sneaker and soda commercials, the sky boxes and corporate contests during timeouts. . . . The new stadiums will seat more people, more comfortably and basketball will have to settle for being a partner with all the corporate sponsors who now invade the local arenas, from Portland to Chicago to Boston, all in the name of progress" (1995, pp. D1, D4). David Noble deplores how the combination of the ideographs technology and progress imprison us, so that "technological development has come to be viewed as an autonomous thing, beyond politics and society, with a destiny of its own which must become our destiny too" (quoted in Leonard, 1993, p. 672). As Opus, in one of the last "Bloom County" comic strips, wistfully explains to the dandelion patch:

> I DON'T SEE WHY YOU GUYS ARE SO UPSET THAT WE'RE LEAVING. I'M NOT UPSET AT ALL. NOT ONE BIT. NOTHING EVER STAYS THE SAME. YOU GUYS SHOULD ESPECIALLY UNDERSTAND THAT. BIGGER AND BETTER THINGS! ONWARD AND UPWARD! PROGRESS IS PROGRESS! AND YOU KNOW WHAT THEY SAY ABOUT PROGRESS: [at this point an Acme Asphalt machine paves over the dandelion patch, leaving Opus to address a highway—with a sign warning "DIP" in the background] IT'S MADE THE WORLD WHAT IT IS TODAY! (Breathed, 1990, p. 84)[4]

The extent to which social control is fundamentally rhetorical and ideographic usages imprison us is most evident when circumstances or antagonisms open potential spaces for change, openings that tend to be encompassed by a "synchronic structure of ideograph clusters constantly reorganizing itself to accommodate specific circumstances while maintaining its fundamental consonance and unity" (McGee, 1980a, p.14). McGee writes that because ideographs "are definitive of the society we have inherited, they are conditions of the society into which each of us is born, material ideas which we must accept to 'belong.' They penalize us in a sense, as much as they protect us, for they prohibit our appreciation of an alternative pattern of meaning" (1980a, p. 9). This sense of ideographs is evident in people's thinking about and reactions to the various environmental crises, a circumstance that has the possibility of changing the meaning of the ideographs progress, nature, and technology or rearticulating the synchronic structure of that cluster of ideographs. Unfortunately, the possibilities of the various environmental crises (antagonisms pointing to the limits of progress) have been circumscribed by economic progress.

For example, in a remarkable article in *Time*'s Planet of the Year issue, the *Time* writer reaches the limit of progress, sees the abyss we're stumbling into, peers into the wilderness, and radically questions one of the underlying tenets of industrialism: that the world exists as a storehouse of resources for humanity to use however it sees fit. But the writer cannot escape the embrace of economic progress and nature, and therefore, even when he begins to question industrialism, he asserts its fundamental presuppositions: "Man must abandon the belief that the natural order is mere stuff to be managed and domesticated, and accept that humans, like other creatures, depend on a web of life that must be disturbed as little as possible. . . . But environmental protection must make economic sense, and development must go hand in hand with preservation" (Linden, 1989, pp. 35, 34). In another instance that demonstrates how the ideographs progress and nature condition us and limit our thinking, *Time*, in a cover story on the spotted owl controversy, labels an environmentalist (an obstructor of progress) who had received death threats "Terrorist in a White Collar" (Seideman, 1990, p. 60). Simultaneously, in another sidebar *Time* bestows upon a logger (a harbinger of progress) who has been logging 44 years and "may have cleared as much of the ancient Northwest forest as any man" the title of "Artist with a 20-lb Saw" (Gup, 1990a, p. 61).

The discourse that we are calling industrialism not only provides the context for environmental discourse in the mainstream media, but it also provides the context for those who are supposedly protecting the environment, such as environmentalists. John C. Sawhill, president of The

Nature Conservancy (and an economist), wrote a column in the *New York Times* arguing for continued protection of endangered species, even if it affects economic growth. Sawhill was responding to then interior secretary Manuel Lujan Jr.'s suggestion that the Endangered Species Act be weakened so that endangered species do not interfere with economic growth. Sawhill's position seems to go beyond the bounds of progress and nature. However, it turns out that Sawhill is arguing to preserve species, what he terms "our 'genetic warehouse,' " for long-term economic growth. "Seemingly useless species have proved to have important, even essential applications" (Sawhill, 1990, p. 23). Other species are understood to have no intrinsic value, just value in terms of what they can do for humanity. "Species diversity is a renewable resource with direct and indirect economic potential" (Sawhill, 1990, p. 23). Ecosystem management is touted as an alternative to the Endangered Species Act because it does not obstruct progress: "Rather than let the life or death of a single species determine the fate of proposed dams and timber projects, this new philosophy of land management was designed to protect broad natural habitats, including the species within them, while still leaving room for humans to exploit the land. Ecosystems, unlike species on the brink of extinction, are elastic enough to take some abuse by humans; they can be preserved without getting in the way of progress" (Cushman, 1995, IV, p. 5). Even the views of environmentalists who are ostensibly opposed to the destruction of the environment in the name of progress tend to be constrained within the ideographic cluster articulated as industrialism. In such a situation, environmentalists become economists and environmentalism becomes economics. "A sound and safe environment and a flourishing economy are two sides of the same coin" (Michael Deland, chairman of the President's Council on Environmental Quality, quoted by Wald, 1990, p. 25). Within the assumptions of progress, environmentalism must make economic sense, so that not only does the environmentalist Vice-President Al Gore extol environmentalism as a path toward economic growth, even such a renowned environmentalist as Barry Commoner argues, "Our task now is to use this experience to show how the strategy of pollution prevention can vastly improve the country's failing program of environmental regulation—and at the same time convert it from a drag on the economy into a major source of economic growth" (Commoner, 1993, p. 25).

READING IMAGE EVENTS

With the rhetorical tactic of image events, radical environmental groups are contesting the hegemonic discourse of industrialism and the received meanings of the ideographs progress, nature, humanity, reason, and technology. Taking advantage of the antagonisms produced within the discourse of industrialism—for example, mass extinctions, deforestation, toxic wastes, nuclear radiation, air pollution, water pollution, and so on—radical environmental groups practice image events not merely to call attention to particular problems but also to challenge the discourse of industrialism and to move the meanings of fundamental ideographs. With such practices, they are attempting to create social movement. As we shall see, the rhetorical tactic of image events works not so much through identification as through disidentification, through the shock or laughter that shatters

> all the familiar landmarks of my thought—*our* thought, the thought that bears the stamp of our age and our geography—breaking up all the ordered surfaces and all the planes with which we are accustomed to tame the wild profusion of existing things, and continuing long afterwords to disturb and threaten with collapse our age-old distinction between the Same and the Other. (Borges, quoted in Foucault, 1973, p. xv)

The impossibility of thinking that a whale or a tree is equivalent to a human, an impossibility that these image events embody, throws into relief the limits of industrialism.

As an example, let us start with a closer reading of Greenpeace's confrontation with the Soviet whaling ship *Vlastny*. The initial image is of the hulking, black whaling ship dwarfing the Zodiacs, which skitter about the ship like waterbugs as it plunges through the ocean in pursuit of a pod of eight sperm whales. The ship also dwarfs the sperm whales it is pursuing. The size of the whaling ship is crucial to the power of the image event. Many have speculated that the current drive to dominate nature is a residue of a time when humans and their cultures were precariously situated within an immense, even infinite, sea of nature (see Nash, 1967/1973; Bateson, 1972; Horkheimer, 1947; Horkheimer and Adorno, 1972). In this relation, nature was a threatening, powerful force that dominated humans. Whales, as the largest animals ever, symbolize the sublime power of nature. They are the monsters of the deep, the leviathans of yore. However, with modern industrial technology humans have reversed the ancient nature/culture dichotomy. This reversal is symbolized in the relation of the *Vlastny* to the whales. Far from the days of *Moby Dick*, the *Vlastny* towers over the whales and slaughters them with ease. For the whalers, whaling is now about as dangerous as trout fishing. The whaling ships are the new leviathans of the deep. Whales, no longer the monsters of legend, are reduced to a resource. In

Like the whales, Greenpeace activists in a Zodiac are dwarfed by a whaling factory ship and a harpoon ship.

this reduction is read the domination of nature. The technological sublime has replaced sublime nature.

Whales and nature, then, are no longer powerful forces threatening people but entities at the mercy of humans and their technology and in need of protection. Greenpeace was aware of the power of this image. As original Greenpeace activist Robert Hunter explains:

> As a newsman, I knew we had achieved our immediate goal. Soon, images would be going out into hundreds of millions of minds around the world, a completely new set of basic images about whaling. Instead of small boats and giant whales, giant boats and small whales; instead of courage killing whales, courage saving whales; David had become Goliath, Goliath was now David; if the mythology of Moby Dick and Captain Ahab had dominated human consciousness about Leviathan for over a century, a whole new age was in the making. Nothing less than a historic turning point seemed to have occurred. From the purely strategic point of view of a media campaign aimed at changing human consciousness, there was little more that we could hope to achieve. (1979, p. 229)

A second key image that works rhetorically to contest the dominant understanding of humanity, nature, and humanity-nature relations occurs when the whaler fires the harpoon, narrowly missing the Zodiac bobbing on the ocean swells before striking the whale. Quite clearly, the Greenpeacers in the Zodiacs, human beings, are risking their lives for animals. This is an almost incomprehensible act in a modern, humanist, secular culture. Over the past 400 years in Western culture God and nature have been displaced in favor of narratives, such as the Enlightenment, scientism, capitalism, socialism, and what I have been referring to as industrialism, that place human reason and humans at the center (see Bateson, 1972; Berman, 1984; Capra, 1983; Foucault, 1973; S. Griffin, 1978; Glacken, 1967; Oelschlaeger, 1991). Humans risking their lives for animals shakes the a priori anthropocentric assumption of these narratives, breaks the Great Chain of Being, and disobeys the command in Genesis: "Be fruitful and multiply, and replenish the earth, and subdue it: and have dominion over the fish of the sea, and over the fowl of the air, and over every living thing that moveth upon the earth" (Bible, Gen. 1:28 King James Version).

While lowering the position of humans in the hierarchy, by risking their lives for the whales the Greenpeacers simultaneously challenge the understanding of animals and nature as mere machines or matter in motion, a storehouse of resources for humans to exploit. These notions are the products of a centuries-long process that Berman felicitously calls "the disenchantment of nature." In short, by placing themselves in the line of fire of a harpoon cannon Greenpeace challenges the anthropocentrism of Western culture and proffers the humble thought that other animals have a right to live and have intrinsic value, not merely economic value. Defining whales as so many gallons of oil and pounds of blubber is too paltry to encompass the richness of these large-brained mammals. In this image event, then, Greenpeace challenges definitions of whales (and by extension nature) as either monsters or economic resources and instead offers whales as coinhabitants of Planet Earth and fellow intelligent mammals. It is worth noting that although Greenpeace's early image events were mind bombs that encouraged disidentifications, they tended to intervene on behalf of mammals that offered potential identifications on some level—from the intelligence of whales and dolphins to the furry cuteness of baby harp seals.

In sum, the image of Greenpeacers in rubber dinghies steering between whaling ships and whales is an encapsulated rhetorical and philosophical statement challenging the anthropocentric position granting humans dominion over all living creatures and implicitly offering biocentrism as an alternative. By arguing against reducing animals to economic resources and instead proposing that animals have intrinsic value and inalienable rights, Greenpeace contests the linking of economic progress with nature as a storehouse of resources, thus challenging the discourse of industrialism that warrants the use of technology to exploit nature in the name of progress.

The image events of Earth First! also interrogate the fundamental ideographs of industrialism while contesting the actions such ideographs warrant. For analysis, let us look at the protester sitting on a platform 100 feet up an old-growth tree and the protester buried up to his neck in a logging road (see the picture on page 125). Akin to the Greenpeace image event, what is striking about both of these images is the utter vulnerability of the protesters as they intervene on behalf of nature. Once again, humans are risking their lives for nature. Perhaps in identification with the form of nature that they are attempting to save, trees, both protesters have rendered themselves relatively immobile. The

An Earth First! activist risks his life ascending to a tree-sit platform precariously perched high in a giant tree.

Earth First! activist on the 8-by-4 platform 100 feet up the tree is help-less if the loggers decide to cut the tree despite the protester's presence (this has happened). The Earth First!er buried up to his neck in the road is utterly helpless. He is exposed not only to the potential anger of log-gers or law enforcement officers, but to the torturous immobility of not being able to use his hands, whether to swat away a mosquito or to scratch an itch. Indeed, his position evokes old pirate and cowboy–Indi-an movies, where victims would be buried on the beach and left to the tide or buried in the desert, covered with honey, and left to the ants.

In performing these image events, the activists translate their hu-manist bodies into ecocentric bodies. Perched high in the Douglas fir, the protester sees the world from the tree's point of view and "becomes" the tree. Rendered relatively immobile, his movements are limited to the swaying of the tree. The protester, like the tree, depends on nourish-ment to come to him. Finally, their fates are entwined as the protester depends on the tree for support and shelter while the tree depends on the protester's presence to forestall the chainsaw. This mutual depen-dence is particularly clear in the case of Julia "Butterfly" Hill, who has lived in a 1,000-year-old redwood, "Luna," for over a year (since De-cember 10, 1997). She has told of how the tree sheltered her during El Niño and some of the worst storms in California history. Her presence, meanwhile, has stopped Pacific Lumber from killing Luna. In the road blockade, the protester buried in the earth becomes the earth. He adopts a ground level view of the world. People and equipment tower over him. He is immobile and must be spoon fed. But his vantage point allows him to speak for the earth: "Defending what's left of the wilder-ness, defending what's left of the world." In clinging to treetops and em-bedding themselves in the earth, the Earth First! protesters both literally perform and symbolically enact humanity's connection to nature. In dis-lodging the blinders of a human-centered worldview, the protesters bring into being an ecocentric perspective. In identifying with the tree and the earth, the protesters invite viewers to identify with the natural world.

As the protester buried in the road speaks, the camera zooms in on him. Technology brings his face and the face of the tree-sitter into my world. Their faces confront me, compel my attention. "A face turned to us is an appeal made to us, a demand put on us . . . there lies the force of an imperative that touches us, caught sight of wherever we see a face turned to us" (Lingis, 1994, p. 167). The weary face of the tree-sitter and the bespectacled, bearded, smiling face popping out of the road tes-tify to their thoughtfulness, resolution mixed with resignation, and hu-manity. In my encounter with these faces, I concur with Lingis: "I find all that I am put into question by the exactions and exigencies of the other. In the face of another, the question of truth is out on each proposition of which my discourse is made, the question of justice put on each move and gesture of my exposed life" (1994, p. 173). The imperatives of these faces call to us and call us to account. They call us to account for proposing an anthropocentric worldview that reduces the rest of the world to a storehouse of resources. They call us to account for industrial practices that destroy a natural world so intimately connected to their bodies, our bodies.

With these image events, then, Earth First! has extended Greenpeace's critique of anthropocentrism. Their choice of trees and ecosystems (e.g., old-growth forests) instead of conscious mammals suggests a move from biocentrism to ecocentrism, a focus not merely on life but on natural systems. The image events of Earth First!, then, contest the possibility of property and the definition of the land as a resource, and instead suggest that biodiversity has value in itself. Progress, then, is not the increasing production of goods through the technological exploitation of nature as a storehouse of resources, but rather the recognition of the intrinsic value and fundamental importance of ecosystems and the need for humans to live within limits as a part of larger ecosystems.

The image events of environmental justice groups in some ways differ significantly from the previously discussed image events, which reflects their local community base. However, they also significantly challenge the discourse of industrialism, as seen in the Allegany County bridge protest where six grandparents chained themselves to a bridge in order to block a commission seeking to site a low-level radioactive waste dump.

Unlike the image events of Greenpeace and Earth First!, the bridge sit-in is not strikingly dangerous, though the protesters court arrest and a possible violent backlash as they seek to protect their community's land. Instead, what is most striking about this image event are the identity of the protesters and the locale of the protest.

Both the Greenpeace and the Earth First! activists fit the stereotype of the young, unkempt, unruly, hippielike, quasi-professional protester so easily branded as outside the mythical mainstream. In the bridge sit-in, however, we are confronted by six white-haired local grandparents huddled in blankets and braving the elements in order to defend their community from bureaucratic outsiders in cahoots with big business (the nuclear power industry). With their very presence, their body rhetoric, the grandparents expose the lie that progress benefits a universal humanity, showing to the contrary that it is ordinary people (us) who are often forced to bear the costs of progress.

The locale of their image event radicalizes the environmental justice activists' critique of industrialism. Both Greenpeace and Earth

First! venture into a nature "out there" to confront the extractive forays into pristine nature of the resource industries. In doing so they can be read as devaluing culture or humanity and revaluing nature. This is important work but it leaves intact a culture/nature hierarchy, even if it is reversed. The bridge sit-in, on the other hand, is placed in inhabited land, a nature that includes people. This is a deconstructive move that not merely reverses but displaces the culture/nature dichotomy, thus contesting the Cartesian contention that there is an ontological divide between the human and the nonhuman. Nature is not a machine or an object to humanity's subject, a homogenous, universalized space. Rather, humans are always embedded in place (not space) in a particular relationship that is coconstitutive of both the identities of the land and the people. Community includes not only people but also animals, plants, waters, and soils. In living a notion of community akin to ecologist Leopold's land ethic (Leopold, 1949/1968, p. 204), they are answering Leopold's call for "the extension of the social conscience from people to land" (1949/1968, p. 209), thus shaking the a priori anthropocentric assumption of Western culture. In its place ACNag implicitly presents an ecocentric perspective that contests the reductionistic definitions of land (and nature) as merely property and resource and offers a moral and practical perspective anchored in the land ethic: "A thing is right when it tends to preserve the integrity, stability, and beauty of the biotic community. It is wrong when it tends otherwise" (Leopold, 1949/1968, pp. 224–225). From this perspective, then, progress is not the increasing production of goods through the technological exploitation of nature as a storehouse of resources and receptacle of wastes, but rather the recognition of the intrinsic value and fundamental importance of ecosystems and the need for humans to live within limits as a part of the land and nature.

With their image events, then, environmental justice groups are fighting an articulation of industrialism that understands progress (economic) as linked to the domination of nature, for the domination of nature involves not only the domination of external nature but the domination of other humans and the domination of one's own inner nature (Horkheimer, 1947, 1972; Horkheimer and Adorno, 1972; Leiss, 1972). As Larry Wilson, a community leader from Yellow Creek, Kentucky, explains, "The system was invented by the people who are poisoning us. The rules say they get to argue over how much cyanide they can put in our coffee, how much poison they can put out before they have to take responsibility for it. That's not a system we can ever win in" (quoted in Greider, 1992, p. 166). In this system certain groups of people (minorities, members of the working class) get abused as a natural resource (labor power) and polluted as nature. For example, the Clean Air Act allows 2.7 billion pounds of toxic chemicals to be spewed into the atmosphere every year (Easterbrook, 1989, p. 28). Environmental justice groups want to move the meanings of progress and nature so that governments are more concerned with people and the environments they are embedded in and less eager to greet corporations with open pockets and closed eyes, as does Kentucky with road signs that announce, "Kentucky is OPEN FOR BUSINESS" (Van Gelder, 1992, p. 64), or Harrisburg, Virginia, with signs stating, "Harrisburg is a certified business location."

In sum, the image events of Greenpeace, Earth First!, and environmental justice groups are a sustained critique of the articulation of humanity, reason, technology, nature, and progress in the discourse of industrialism. In addition, they challenge the meanings of particular ideographs. Although this discussion has focused on progress and nature, they also interrogate the accepted universalization of humanity as "rational man," the Cartesian subject. In the image events discussed, we witness people acting passionately ("irrationally") on behalf of nature and place, commitments that owe as much to love and emotional connections as they do to instrumental reason. Indeed, often these image events are refuting the results of a scientific rationality that through the methods of cost–benefit analysis and risk assessment sanctions environmental destruction, rising cancer rates, extinctions, and deaths per thousand in exchange for profits. Greenpeace has and continues to challenge the "scientific whaling" of Japan, Iceland, and the former Soviet Union. Earth First! is questioning the very possibility of "science" (a neutral universal practice based on reason) as it condemns the science of the U.S. Forestry Service that recommends clear-cutting and other practices that most clearly benefit the timber, oil, and mining industries. The AC-Nag bridge-sitters, in blocking the low-level nuclear waste dump Siting Commission (composed of scientists), are refusing to accept the scientific assurances of distant experts that the dump will be safe.

Image events of radical environmental groups have tended not to be recognized as rhetorical acts working for social movement not only because they fall outside traditional definitions of rhetoric and social movement, but also because they do not fall within the modernist frame of politics. The tactical image events of radical environmental groups are not primarily directed toward legislation, electoral politics, the distribution of material goods, and the expansion of institutional political rights, nor even a Marxist revolution. Rather, radical environmental groups contest social norms and deconstruct the established naming of the world.

This shift is a response to two factors. The first is the realization that environmental problems are hardly recognizable and certainly not solvable within the existing framework of industrialism, wherein all problems are subject to the economic calculus of progress. This is abundantly clear in third-wave environmentalism, which construes industrial capitalism, once understood as a cause of environmental problems, as the solution. The second is the recognition that the discourse of industrialism has infiltrated all spheres: political, governmental, economic, and social. Consequently, since industrialism is not concentrated in a centralized seat of power, revolutionary overthrow of such a seat of power is not possible. Further, political responses, if they are to be effective, cannot be delimited to a formal sphere of politics. This is a recognition of power not as monumental but fluid, power not only as unitary, centralized, and institutionalized, but also as diffuse, capillary, and everywhere (Foucault, 1978, pp. 92–98; Patton, 1990, pp. 124–126). If power is understood as emanating and "exercised from innumerable points, in the interplay of nonegalitarian and mobile relations" (Foucault, 1978, p. 94), resistances are necessarily multiple and dispersed. The power of King Coal in Kentucky is in the State House and in the shacks of Appalachia and in the economic and architectural structures of coal towns and in the generations of coal miners and in the memories of dead coal miners and in the futures of school children and in the scarred landscape. The resistance of KFTC cannot be localized in the State House, but must exist in multiple localities—the State House and coal towns and private homes and cities and garbage dumps and county fairs and churches and zoning meetings and coffee shops. To more fully account for the rhetorical force of image events, we need to reconsider what counts as politics in our present social–historical moment.

MEDITATION II

There are thus two interpretations of interpretation, of structure, of sign, of play. The one seeks to decipher, dreams of deciphering a truth or an origin which escapes play and the order of the sign, and which lives the necessity of interpretation as an exile. The other, which is no longer turned toward the origin, affirms play . . . the Nietzschean *affirmation,* that is the joyous affirmation of the play of the world and of the innocence of becoming, the affirmation of a world of signs without fault, without truth, and without origin which is offered to an active interpretation. *This affirmation then determines the noncenter otherwise than as loss of the center.*

> —Jacques Derrida, "Structure, Sign, and Play in the Discourse of the Human Sciences"

CHAPTER 4

THE POSSIBILITIES OF NATURE IN A POSTMODERN AGE

The Case of Environmental Justice Groups

In *The Condition of Postmodernity* David Harvey (1989) recognizes postmodernism as "a sea-change in cultural as well as in political–economic practices" and devotes much of his book to arguing compellingly that, rather than another fad in academia, postmodernism names a further transformation in capitalism (akin to Jameson's "late capitalism"), evidenced by a shift from production capital to fictitious capital; Fordism to flexible accumulation; state power to financial power; mechanical production to electronic reproduction; specialized workers and trade unions to flexible workers and temporary contracts; and industrialization to deindustrialization (pp. 340–341). Yet despite the seriousness and thoroughness with which Harvey treats postmodernism as a historical phenomenon, at times he can't stop himself from reverting to a moral stance and blasting postmodernism as a reveling in diversity, simulation, and fragmentation that makes politics impossible (pp. 116–117).

Although it may be comforting to dismiss Harvey's reaction as the knee-jerk reflex of an orthodox Marxist, certainly his sentiment is not restricted to Marxists. According to their critics,[1] postmodernists, lacking the unifying principle of a foundational premise (be it the humanist subject, Reason, or the economic mode of production and the laws of historical materialism), are reduced to making silly gestures (Lyotard), celebrating the reactionary effects of market capitalism (Baudrillard), championing alternative lifestyle choices (Foucault), or lapsing into political silence (Derrida). Devoid of a guiding metanarrative, new social movements are condemned to practicing the fragmentary politics of special interest groups (Harvey, 1989, p. 302).

In communication circles, these criticisms are echoed by the likes of Dana Cloud, who worries that the postmodern project leads to "the evacuation of the critical project" in favor of "the aestheticization and depolitization of political struggle" (1994, pp. 159, 157); Roderick Hart, who condemns a postmodern politics characterized by a "panoply of skepticisms" (1994, p. 164); and G. Thomas Goodnight (1995, pp. 283–286), who decries the postmodern play of mimicry and *ilinx* (a whirlpool signifying dizziness and confusion), which reduces the "real thing" to nothing but "the interminably recycled moments of detachment, disavowal, and cynicism flowing from and into a mediating code of cultural skepticism" (1995, p. 285). Taking seriously these fears of a postmodern planet, I read them as clues to the central question of our time: What are the possibilities of politics in a postmodern age?

Harvey's answer is that postmodernism's penchant for deconstructing foundations and metanarratives and fetishizing locality and place lead to an incoherent politics that isolates and disempowers local resistances while aiding global corporate capitalism. Is it possible, however, that deconstructing transcendental foundations, inhabiting places, and living with incoherence offer hope for a radical democratic politics? I will attempt to answer this question by exploring the political possibilities offered by the status of nature in a postmodern age. In particular, I will argue that radical environmental groups (with a focus in this chapter on environmental justice groups), through their unorthodox rhetorical tactics, question the modern grand narrative of industrial progress, seek to rearticulate identities, work toward reinventing "nature," open new possibilities for humanity-nature and human–human relations, and break with conventional politics and rhetoric through the practice of a radical form of participatory democracy, thus enacting the political and rhetorical possibilities of a postmodern age. Their effects will be measured in relief to the modern strategies of mainstream environmental groups.[2]

This description of the tactics of radical environmental groups as concerned with identity formations, namings of the world, and social constructions of reality, reads such groups as engaged in discourse politics. Discourse politics becomes a crucial arena of contestation at a time when power is clearly repressive and productive, decentralized and capillary. Such an understanding of power expands politics beyond the institutional realm, requiring a concomitant expansion of resistance. Politics is expanded so that "a matter is 'political' if it is contested across a range of different discursive arenas and among a range of different discourse publics" (Fraser, 1989, p. 167).

The question "What are the possibilities of politics in a postmodern age?," then, can also be read as "What are the possibilities of discourse politics?" As a discourse theory, the rhetorical theory of social movement should be able to answer that question and help explain how social identities and social groups are formed and transformed, how cultural hegemony is secured and contested, and how social change is effected. Chapter 2 worked to answer the latter questions. The primary task of this chapter is to explore how the disruption of the ideograph nature, a foundational element of the articulation of industrialism, opens up possibilities for politics.

MODERNISM, NATURE, AND POSTMODERNISM

Over 20 years ago prescient media theorists recognized that technological advances had destabilized the meaning of nature. Marshall McLuhan observed: "When Sputnik (1957) went around the planet, the planet became programmable content, and thus became an art form. Ecology was born, and Nature was obsolesced" (1977, p. 80). His disciple, Jean Baudrillard, concurred: "The great signified, the great referent Nature, is dead. . . . If one speaks of environment, it is because it has already ceased to exist. To speak of ecology is to attest to the death and total abstraction of nature" (1972/1981, p. 202). Their counterintuitive observation is now echoed by a chorus, in particular by Bill McKibben in the bestseller *The End of Nature*: "There's no such thing as nature anymore. . . . We have killed off nature—that world entirely independent of us which was here before we arrived and which encircled and supported our human society" (1989, pp. 89, 86)

If nature is commonly understood as the nonhuman, the wilderness untouched and untainted by humans, a storehouse of resources that circumscribes and sustains humanity, the reference point for civilization, these writers announce the death of nature, or at least the crisis of nature. The death of nature is both material and conceptual. Material conditions have changed. Humans have penetrated every corner of the earth, and human activities have altered the global temperature, weather patterns, and the very composition of the atmosphere. Advances in science and technology have undermined the boundaries between human and animal, organism and machine, and the physical and the nonphysical (Haraway, 1991, pp. 151–153). Conceptually, it is difficult to conceive of nature as an immense, primal power that surrounds, sustains, and threatens culture or simply as "that which is not culture" when the boundary between nature and culture is so thoroughly breached and we are witnessing the assimilation/consumption of nature by culture, what Jameson calls "an immense and historically original acculturation of the Real" (1991, p. x). The nature in crisis is modern nature, and this crisis is part of the crisis of modernism and a manifestation of postmodernism.

As discussed in Chapter 3, modernism is characterized by a belief in the grand narrative of progress or industrialism: humanity, by dominating nature through the use of reason and technology, will achieve progress.[3] To put it in shorthand, modernism is "the conjunction of Bacon's, Descartes's, and Locke's programs, the fusion of the domination of nature, the primacy of method, and the sovereignty of the individual" (Borgmann, 1992, p. 42). From this perspective, it is clear how the crisis of nature is related to the crisis of reason (instrumental reason) and the crisis of the subject. As Haraway points out, the crisis of nature is one of the causes of the crisis of reason: "The certainty of what counts as nature—a source of insight and promise of innocence—is undermined, probably fatally. The transcendent authorization of interpretation is lost, and with it the ontology grounding 'Western' epistemology" (1991, pp. 152–153). These crises are characteristics of postmodernism; indeed, the crisis of nature is central to postmodernism. Jameson goes as far as to argue that "postmodernism is what you have when the modernization process is complete and nature is gone for good" (1991, p. ix). At the very least, in postmodernism nature is no longer considered the ground and horizon of culture. Indeed, there is the inversion of the framing and bounding of culture by nature.

While heeding Nietzsche's warning that "all concepts in which an entire process is semiotically concentrated elude definition" (quoted in Calinescu, 1987, p. 310), I want to suggest that postmodernism can be characterized in part by the conjunction of the following elements: a decentering of the subject as origin, end, and arbiter of theory and practice; a destabilization or fragmentation of all kinds of identity; a lack of belief in any foundation, totality, transcendental signified, or grand narrative; a move from the domination of sameness to the recognition of difference; a generalized awareness of limits, particularly the limits of reason (Laclau, 1990, p. 3); a valorization of the local in the face of inexorable globalism; a change in material conditions, including the disappearance of nature as the great referent that ontologically grounds Western epistemology; time–space compression; the displacement of nation-states by transnational corporations; and the rise of both image politics and micropolitics.[4]

These characteristics, which together constitute the structure of feeling known as postmodernism, especially the displacement of the subject, reason, and nature, can lead to pessimism and conservative retrenchment, but they also can be the occasion for radical optimism. For example, the dislocation of the modernist concept of nature as a storehouse of resources opens up the possibility of critiquing the domination of nature and rearticulating human–nature relations. The dislocation of the subject, reason, and nature open, as Ernesto Laclau notes, "unprecedented opportunities for a radical critique of all forms of domination, as well as for the formulation of liberation projects hitherto restrained by the rationalist 'dictatorship' of the Enlightenment" (1990, p. 4).

Of course, for those active in environmental movements, the death of nature as the object of environmentalism may seem less an occasion for radical optimism and more a cause for political confusion, if not despair.[5] The loss of a stable, essential foundation for a political movement has far-reaching effects. In the next section I want to explore what postmodern feminisms can teach environmentalists in a postmodern age.

FEMINISMS AND ENVIRONMENTALISMS: AFTER "WOMAN" AND "NATURE"

Many feminisms and environmentalisms have been animated by essentializing visions of "woman" and "nature," respectively. These tendencies to essentialize foundations converge in radical versions of ecofeminism,[6] in which not only are claims made that "there are important connections between the oppression of women and the oppression of nature" and that "understanding the nature of these connections is necessary to any adequate understanding of the oppression of women and the oppression of nature" (Warren, 1987, p. 4), but, moreover, it is claimed "that the earth is fundamentally feminine rather than masculine . . . that nature is intrinsically feminine" (Oelschlaeger, 1991, pp. 309, 311). In such claims, the observation that "women and nature have an age-old association—an affiliation that has persisted throughout culture, language, and history" (Merchant, 1991, p. 258) is reduced to an essential connection fundamental to the meanings of "woman" and "nature." For example, such a perspective leads Marti Kheel to argue that "it is out of women's unique, felt sense of connection to the natural world that an ecofeminist philosophy must be forged" (1990, p. 137), a unique connection fostered through Charlene's Spretnak's "'body parables'—'reclaimed menstruation, orgasm, pregnancy, natural childbirth and motherhood'" (Spretnak quoted in Kheel, 1990, p. 137).

Such essentializing claims have been criticized as making anatomy into destiny, falling prey to patriarchal sex-role stereotyping, universalizing culture-specific conceptual categories, and reinforcing the culture–nature binarism central to the domination of nature (Li, 1993;

Merchant, 1990; Oelschlaeger, 1991; Quinby, 1990; Stabile, 1994; Warren, 1987). This ecofeminist version of difference feminism perpetuates a modern nature that limits what can be considered an environment deserving of protection, narrows what can be counted as environmental politics, and blocks necessary coalitions across gender, race, and class lines. Perpetuating modern nature also reproduces the other problematic aspects of modernism, including the Cartesian subject, scientific reason as the universal method to foundational Truth, and linear progress. As Lee Quinby puts it, "Essentialist ecofeminism speaks of a monovocal subject, Woman; of a pure essence, Femininity; of a fixed place, Nature; of a deterministic system, Holism; and of a static materiality, the Body. Writings that give homage to the Goddess as a symbol of *the way things really are* . . . echo masculinist prescriptions about the way things 'really' are—and must always be" (1990, p. 126). Even more ominous is that by "accepting (indeed, celebrating) the logic that posits a special connection between women and nature, ecofeminist philosophies maintain hazardous ties with anti-feminist, anti-environmentalist, and racist conservative ideologies" (Stabile, 1994, p. 61; Merchant, 1990, p. 102, echoes Stabile). Postmodern feminist Judith Butler is a particularly powerful critic of difference or essentialist feminism. Her critique and her advice for political action offer directions for environmentalists.

In *Gender Trouble* (1990) and in "Contingent Foundations: Feminism and the Question of 'Postmodernism,'" (1992) Butler's task is to subvert and destabilize foundations that are posited as origin and cause but are "in fact the *effects* of institutions, practices, discourses" (1990, p. ix). To assume and require an unquestioned and unquestionable foundation for a politics is an authoritarian move used to silence political possibilities. The necessary response to such a move "is to interrogate what the theoretical move that establishes foundations *authorizes,* and what precisely it excludes or forecloses" (Butler, 1992, p. 7). For example, to establish nature as a realm apart from humanity as the foundation of the environmental movement—the protection of such a nature being the reason for the environmental movement's existence—reinforces nature as object to humanity's subject, authorizes treating nature as a collection of objects or potential storehouse of resources, excludes other meanings and concepts of nature, and forecloses possible human–nature relations not based on an ethic of domination. To question the political construction of such a meaning of nature is to remove the foundational ideograph nature[7] from the epistemologically given and to make it a site of political struggle. Specifically, in this case, to deconstruct nature as that which is external (a storehouse of resources) to humanity and culture raises questions as to who constructed this meaning and why and upon what exclusions this foundational ideograph is founded. Further, such a deconstruction frees nature from certain Western ontologies and opens it up to possibilities and redeployments that have not previously been authorized and that may serve alternative political aims (Butler, 1992, pp. 15–17). Butler notes, "It is this movement of interrogating that ruse of authority that seeks to close itself off from contest that is, in my view, at the heart of any radical political project. Inasmuch as poststructuralism offers a mode of critique that effects this contestation of the foundationalist move, it can be used as a part of such a radical agenda" (1992, p. 8).

In short, Butler's critique suggests that environmentalists ought not to mourn or revive "nature"—"that world entirely independent of us which was here before we arrived and which encircled and supported our human society" (McKibben, 1989, p. 96). To cling to such a stable, essentialized meaning of nature (even if to revalue such a nature) is to unwittingly reify human–nature relations in terms of dominance and to support systems of power that exclude other possible notions of nature as unintelligible and impossible. Wendell Berry observes that "if, even as conservationists, we see the human and the natural economies as necessarily opposite or opposed, we subscribe to the very opposition that threatens to destroy them both" (1987, p. 18).

Following Butler's reasoning, the task for environmental activists is to promote the detachment of the ideograph nature from any foundational meaning and, instead, to understand nature as a culturally constructed ideograph in the open social field of discursive politics. When nature is thus understood, environmental movements become key sites for the "invention and reinvention of nature—perhaps the most central arena of hope, oppression, and contestation for inhabitants of the planet earth in our times" and constructions of nature become understood "as a crucial cultural process for people who need and hope to live in a world less riddled by the dominations of race, colonialism, class, gender, and sexuality" (Haraway, 1991, pp. 1, 2).

STRATEGY AT THE CENTER

Recognition of nature as a contested ideograph is evident in both environmental theory and practice. Deep ecologists, social ecologists, ecofeminists, preservationists, and others have put forth a proliferation of natures that reveal the constructedness of nature, thus subverting the modernist discourse of industrialism that is partially founded on the stable identity of nature as a realm apart from humans to be used as a storehouse of resources. The most potent of these notions of nature— nature as miracle (Evernden, 1989) and nature as coyote or coding trickster (Haraway, 1991)—promote a nature that is actively problematic, eluding stable categorization and causing trouble.

In practice, over the past 15 years there has been a dramatic shift from the strategies of national mainstream environmental groups to the tactics of independent, grassroots environmental groups. Since the late 1970s and early 1980s the strategy of mainstream environmental groups has been to present a unified front on issues in order to protect an anthropocentric vision of nature as an aesthetic and recreational as well as economic resource within an industrial framework. Essential to this strategy was the establishment of a center of power (Washington, DC) as a base from which to protect and to build on past victories through legislative lobbying and national direct-mail fundraising. To that end, in 1981 mainstream environmental groups established quarterly meetings of the Group of Ten[8] modeled on the CEO meetings of the Business Roundtable in order to more efficiently institutionalize environmentalism (Gottlieb, 1993, pp. 117–124). When used in the sense defined by Michel de Certeau in *The Practice of Everyday Life*, the term "strategy" is descriptive both of the actions of the mainstream environmental groups and of the groups themselves:

> I call a *strategy* the calculation (or manipulation) of power relationships that become possible as soon as a subject with will and power (a business, an army, a city, a scientific institution) can be isolated. It postulates a *place* that can be delimited as its *own* and serve as the base from which relations with an *exteriority* composed of targets or threats (customers or competitors, enemies, the country surrounding the city, objectives and objects of research, etc.) can be managed. As in management, every "strategic" rationalization seeks first of all to distinguish its "own" place, that is, the place of its own power and will, from an "environment." . . . A Cartesian attitude. . . . It is also the typical attitude of modern science, politics, and military strategy. (1988, pp. 35–36)

The strategy of the mainstream environmental groups, then, resulted in organizations isolated from their environment (which is particularly worrisome for environmental groups), devoted to the preservation and expansion of their own power, dependent on fundraising to support a bloated bureaucracy, and more cautious in using their increased power to criticize the political and economic establishment since their power depends on their working within the context of the political, economic, and social institutions of society. Former Sierra Club director Michael McCloskey's boast that "we may be 'reformist' and all, but we know how to work within the context of the basic institutions of the society" (quoted in Sale, 1986, pp. 32–33) is, for critics, precisely the problem as mainstream environmental groups become dependent on business managers and lawyers to run their organizations; corporate polluters like Exxon, Chevron, and McDonald's to fund their organizations; and the corporate elite to serve on their boards of directors (Beasley, 1991; Dowie, 1995; Gottlieb, 1993; Pell, 1990; Sale, 1986, 1993). The necessities of strategy in a way trap these groups: "Power is bound by its very visibility" (de Certeau, 1988, p. 37).

It is also worth noting that de Certeau writes that strategy is characteristic of modern institutions. So mainstream environmental groups, which arose in response to the problems of modernism, have adapted the form and practices of the modern institutions they are purportedly criticizing. Such practices include a reliance on scientific expertise, legal acumen, and legislative lobbying. This strategy, then, commits them to modernist concepts of the subject, nature, and humanity–nature relations that are at the root of the environmental crisis. Grassroots environmental justice activist Richard Moore concludes, "They have become the very enemy they originally set out to fight" (Beasley, 1991, p. 42).

This cooptation is dramatically illustrated in the revolving-door relationships among mainstream environmental groups, government, and industry (Gottlieb, 1993, pp. 130–131; Sale, 1993, pp. 54–55). Jay Hair, former longtime president of the National Wildlife Federation, now does public relations for Plum Creek Timber (Cockburn, 1997, p. 10). Wilderness Society representative Walter Minnick also happens to be the CEO of the multinational timber company TJ International. At the 1993 Portland Forestry Summit, he told President Clinton: "Essentially what we need the government to do is get out of the way, let the market system work, get some certainty into the west side timber supply because we don't know whether to build another plant here or go to Canada" (quoted in Cockburn, 1995c, p. 300). The similarities between mainstream environmental groups and corporations also extend to an emphasis on the same bottom line: "Each of the organizations, including Audubon, also strongly emphasized that CEOs had a fiduciary or corporate responsibility that measured group success by the bottom line" (Gottlieb, 1993, p. 153).

The constraints of strategy are manifested in the rhetoric and practices of mainstream environmental groups. Many of these groups put out glossy magazines littered with car and oil company advertisements. John Sawhill, president of the Nature Conservancy, appears in an advertisement for General Motors touting their shared goal: "safeguarding the environment without destroying jobs or businesses. That's a goal General Motors shares" (quoted in an advertisement in *Audubon*, 1995, p. 69). In the name of third-wave environmentalism, the mainstream groups advocate cooperation with corporations. To that end, besides dialoguing with industry and sharing board of directors members, the Environmental Defense Fund pushes pollution credits while the Natural Resources Defense Council and the National Wildlife Federation advocate unfettered market forces as the way to ensure environmental protection (Cockburn and St. Clair, 1994, p. 764; Dowie, 1995, pp. 105–124).

To accommodate their allies in government and industry, mainstream groups often take anti-environmental stands. In 1990, the Sierra Club agreed to a doubling of timber sales in an Interior Department appropriation bill (Sale, 1993, p. 93). The Sierra Club has opposed the Zero Cut campaign calling for no logging on public land (Dowie, 1995, pp. 216–219) and did not oppose logging or even clear-cutting in national forests until 1996. Even this belated position is thrown into question by the election in May 1998 of the latest Sierra Club president, Chuck McGrady. McGrady is a Republican and corporate lawyer known as a "pragmatist . . . [who] opposed the club's 1996 vote to oppose all logging on national forests, 'contending that the club would appear extremist'" (Greenwire, 1998).

The most egregious recent examples of how the strategy of mainstream groups ensconces them within the discourse of industrialism come to us courtesy of the Wilderness Society. Founded in the 1930s by the legendary Leopold and Robert Marshall with the mission to "preserve the freedom of the wilderness," recently the Wilderness Society requested that a judge overturn an injunction halting logging in Idaho's national forests that had been issued at the request of the Wilderness Society. By bowing to pressure from funders, Idaho's congressional delegation, and timber companies, the Wilderness Society enabled clear-cutting to continue in Cove Mallard, America's largest roadless area outside of Alaska (Cockburn, 1995b, 1995d). Consistent with the restraints of strategy, Wilderness Society president Jon Roush declared, "We shouldn't let people cast us in the position of being regulatory zealots. We have to seize the center" (quoted in Cockburn, 1995a, p. 15).

At roughly the same time as reversing the injunction, Roush was logging off 80 acres of mature and old-growth forest through clear-cutting and high-grading (cutting the most valuable trees). Roush's Montana ranch borders the Bitterroot National Forest and is within the largest wild ecosystem left in the United States mainland. Roush, who has criticized timber companies that "measure the value of land only in dollars, in board-feet of lumber" (quoted in Cockburn and St. Clair, 1995, p. 558), reaped at least $140,000 from the sale of the 400,000 board-feet of timber to Plum Creek Timber, a company a Wilderness Society advisor had accused of Nazi forestry.[9] In his defense, Roush blamed his wife, citing their separation agreement (Cockburn and St. Clair, 1995; Cockburn, 1995e; Kenworthy, 1995).

These examples point to the ways that the strategy of mainstream groups ensconces them within the discourse of industrialism and limits their possibilities. Granted, mainstream environmental groups have been instrumental in passing important legislation, including the Clean Air, Clean Water, and Endangered Species Acts, as well as weaving environmentalism into the fabric of everyday life (Sale, 1993, p. 96). Their work remains necessary. Yet more than half the population of the United States lives in counties that violate the Clean Air Act; environmental legal defense organizations spend much of their time suing the government to enforce its own laws (Sale, 1993, p. 90; Dowie, 1995, pp. 37–38); corporations often appeal to federal and international authorities to preempt stricter state and local laws (Greider, 1992, pp. 178–182); over 2 billion pounds of toxic chemicals are spewed into the environment legally every year (see Gray, 1998 and the U.S. Government's Toxic Release Inventory); a Friends of the Earth staffer admits that "it's often a matter of one step forward and two steps back" (quoted in Sale, 1993, p. 56); and the mainstream environmental groups promote a third-wave environmentalism that limits environmental problems to market solutions, establishes the right to pollute, and recasts environmentalists as caretakers instead of critics of the market-driven industrial order. In sum, retaining a modern notion of nature and adapting the organizational structure and strategy of modern institutions prevent mainstream environmental groups from fundamentally challenging the industrial exploitation of nature as resource and reduce them to players in the perpetuation of the industrial system.

TACTICS ON THE MARGINS

Recognizing the limits of the strategy of the mainstream environmental groups and the necessity of working in a postmodern age to "protect" something only contingently defined as "nature" would seem to require a shift in forms of political action, a shift from monolithic environmentalism to heterogeneous environmentalisms that lack a centralized organization able to dictate national strategy. The rise of grassroots environmental groups offers new directions for environmentalism that transgress the limits of strategy and suggests the possibilities of radical democracy for countering global industrialism. Journalist Kirkpatrick Sale describes these grassroots groups, comprised mostly of environmental justice and local wilderness protection groups, as an upstart movement that "has no agreed-upon name, nor even much in the way of cohesion or self-identification. It has roots in, and derives its basic tenets from, a confounding variety of ideas and doctrines. Its tactics run the gamut from petitions and letter-writing campaigns to alternative assemblies and even full-scale ecological sabotage" (1986, p. 26).

The following groups are examples of the diversity of this "movement." Earth First!, as previously discussed, is a grassroots yet national disorganization with no official membership whose tactics range from ecotage and civil disobedience protests (Redwood Summer) to going to city council meetings and weeding patches of prairie. The Oregon Natural Resources Council relies on lawsuits and protests to save old-growth forests. The Sea Shepherd Conservation Society uses direct, often violent, action to protect marine animals. Citizens Clearinghouse for Hazardous Wastes helps organize local groups to fight dumps and incinerators.

Just as the concept of strategy describes the mainstream groups' modes of operation, the concept of tactics describes the multifarious grassroots groups' modes of operation. In a dialogue with social ecologist Murray Bookchin, Foreman says, "We need to delay, resist, and thwart the current system using *all* the tools available to us. . . . I believe in monkeywrenching it, thwarting it, helping it to fall on its face by *using its own stored energy against itself*" (quoted in Chase, 1991, pp. 69, 45; emphasis added). Foreman's sense of tactics for local grassroots groups is akin to de Certeau's definition of tactics:

> A *tactic* is a calculated action determined by the absence of a proper locus. . . . The space of a tactic is the space of the other. Thus it must play on and with a terrain imposed on it and organized by the law of a foreign power. . . . It is a maneuver "within the enemy's field of vision" . . . and within enemy territory. It does not, therefore, have the options of planning general strategy and viewing the adversary as a whole within a district, visible, and objectifiable space. It operates in isolated actions, blow by blow. It takes advantage of "opportunities" and depends on them. . . . [It] must accept the chance offerings of the moment, and seize on the wing the possibilities that offer themselves at any given moment. It must vigilantly make use of the cracks that particular conjunctions open in the surveillance of the proprietary powers. It poaches in them. It creates surprises in them. It can be where it is least expected. . . . In short, a tactic is an art of the weak. . . . The weak must continually turn to their own ends forces alien to them. (1988, pp. 36–37, xix)

To flesh out this concept of tactic in relation to environmental groups, let us take a closer look at Kentuckians for the Commonwealth (KFTC) and Allegany County Non-violent Action Group (ACNag).

KFTC is a grassroots environmental justice group working in Appalachia. Among other issues, KFTC fights against the importing and dumping or burning of hazardous waste, which means fighting powerful corporations and politicians used to making decisions "in back rooms, done deals that benefit a few folks" (Harker quoted in Van Gelder, 1992, p. 66) in a state that greets visitors with signs that announce "Kentucky is OPEN FOR BUSINESS" (Van Gelder, 1992, p. 64). KFTC operates without a power base and is mobile. KFTC's Housewives from Hell, Patty Wallace and Ruth Colvin, for example, operate out of their car, the Toxicmobile, which "has logged nearly 120,000 miles in the service of fighting hazardous waste" (Van Gelder, 1992, p. 62). KFTC operates within enemy territory and creates surprises: spying on the comings and goings of garbage trucks in order to force state officials to enforce their own state laws; pushing for county zoning to prevent the building of an incinerator; campaigning at fairs and churches for a referendum against broad form deed strip mining; and, as described in Chapter 1, performing multifarious image events (Van Gelder, 1992, pp. 62–67; Zuercher, 1991, pp. 121, 150–152).

The effectiveness of KFTC's tactics has given "the politicians a glimpse of something they don't have any idea of how to control or use

The members of Kentuckians for the Commonwealth (KFTC) carry the casket of "Kentucky" up the steps to the state capitol. In the rotunda they held a funeral for a Kentucky "buried in waste."

or even get out of the way of'" (Holwerk quoted in Van Gelder, 1992, p. 67).[10] KFTC and other grassroots groups are also saying to "the national organizations that it's not good when a handful of people are making policy. It doesn't matter whether it's national environmental groups or national polluters. We need to start making decisions from the bottom up. Most of these issues are easy to understand and the people who live close by to a problem learn in a hurry" (True, quoted in Van Gelder, 1992, p. 67). Tactics require people close by who can seize on the wing the possibilities that offer themselves at a given moment, possibilities that distant central organizations can never see in time.

The tactics of grassroots groups not only constitute anew what counts as political practice, they also constitute anew what counts as an environmental group—KFTC considers itself a "'citizens social justice group' to reflect KFTC's broader goals and multi-issue nature" and to avoid "being labeled an 'environmental group'" (Zuercher, 1991, p. 136)—and who are environmental activists: "KFTC members are regular working people: farmers and miners, preachers and school teachers, young and old, Black and white. Many come from families that have lived in Kentucky for five or six generations" (Van Gelder, 1992, pp. 62–64). Finally, with their tactics grassroots groups contest nature as a storehouse of resources apart from humans and reconstruct nature as an open site of possibilities: pristine wilderness, reclaimed pasture, drinking water, air, community spaces, the places people inhabit, even people. Opening up the meaning of nature does not guarantee only beneficial relations. However, the recognition of humans as part of nature, as embedded in the environment, suggests an orientation that would mitigate against the abuses encouraged by defining nature as an objectified source of resources. In short, the opening of the ideograph nature as a site of political contestation does not guarantee results, just hope.

Although de Certeau's concepts of strategy and tactic have tended to be deployed unscathed (Nakayama and Krizek, 1995, p. 295; Browne, 1993, p. 467), the foregoing analysis of the tactics of environmental justice groups suggests certain refinements. De Certeau writes that strategy requires a *place* of power from which to control the exterior environment. I think a more apt description would be that strategy requires a *center* of power from which to control *space*. *Place* implies a particular locality of which a person has an intimate knowledge derived from passionate attachment and caring inhabitation, while *space* suggests an impersonal geometrical region known through the rationalized, objective methods of science. The definitions are telling. While place is "a particular portion of space; the portion of space occupied by a person or thing," space is "the unlimited expanse in which all material objects are contained" (*Random House Dictionary*, 1978, pp. 682, 852). I am suggesting, then, that place is a practiced space, which reverses de Certeau's distinction of "*space is a practiced place*" (1988, p. 117, emphasis in original). Yet de Certeau does argue that strategies "privilege spatial relationships" (1988, p. 38) and deploys Merleau-Ponty's distinction between geometrical space and anthropological space (1988, p. 117), which mirrors my distinction between space and place.

In my reformulation, place is the space of tactics. Although environmental justice groups must improvise off of moments in a legal and legislative space controlled by the strategies of proprietary powers, their ability to take advantage of opportunities in time depends on their roots in place. As the tactics of KFTC and ACNag make evident, local knowledge, love of place, and experience of problems are crucial to waging postmodern politics. In sum, de Certeau's conclusion that strategies bet on place while tactics bet on time should be amended to "strategies bet on control of space while tactics bet on time and place."

Although de Certeau writes that tactics are the art of the weak, thus implying that if the weak gain strength they would turn to strategies and the form of institutional organizations, my characterization of the possibilities of postmodern politics argues for the perpetual practice of tactics. Indeed, environmental justice groups' challenge to industrialism rests in part on their refusal to adopt the form of modern institutions that strategy necessitates. They resist the organizational imperatives of self-preservation, hierarchal authority, and division of labor. Instead, environmental justice groups tend to be issue-driven, hyperdecentralized organizations that eschew official leaders, national headquarters, membership lists, dues, bylaws, and boards of directors. The Coalition on West Valley Nuclear Waste (in New York State), for example, meets in Emil Zimmerman's barn: "Its membership amounts to anyone who shows up in Zimmerman's barn on a meeting night. Most of its funding comes from passing around an empty coffee can" (Luoma, 1991, p. 88).

Even more organized environmental justice groups, such as KFTC, are structured in a way that subverts traditional organizational goals while privileging active participation by ordinary people, local knowledge, and passionate commitment to place. KFTC is composed of local chapters that have equal rights and responsibilities within the statewide organization. Chapter members elect local officers and state steering committee representatives. Enacting their belief that everyone has the potential to lead, KFTC ensures multiple leadership through 3-year term limits for steering committee representatives and 2-year term limits for statewide officers (Zuercher, 1991, pp. 156–157, 167–168). The structures of environmental justice groups suggest that the use of tactics and resistance to modern forms of organization can be a choice and an appropriate choice in a postmodern social field, not merely the only option of the weak.

TALKING ABOUT NATURES/ENVIRONMENTS

Importantly, not only in their image events and other tactics but also in their verbal rhetoric grassroots environmental groups consciously deconstruct the modern concept of nature as a realm apart from humanity, an object to humanity's subject, the binary opposite of culture. This is particularly evident in the discourse of the diverse groups that comprise the environmental justice movement, in which environmental justice activists take the word "environment," used by mainstream environmentalists as a synonym for nature, and redefine it in ways that undermine the culture–nature dichotomy that partly founds modernism. Lois Gibbs, former leader of the Love Canal Homeowner's Association and founder of Citizens Clearinghouse for Hazardous Wastes (CCHW), writes:

> Over the past ten years the Movement for Environmental Justice has redefined the word environment. No longer does the media, the general public or our opponents see the environmental movement as one that is focused on open spaces, trees and endangered species alone. They have finally got it! The Environmental Justice Movement is about people and the places they live, work and play. (1993, p. 2)

Activist Scott Douglas from Birmingham, Alabama, expresses a similar position: "The environment is wholistic. Our movement has always talked about how you can't limit the environment to trees, rabbits, squirrels, and owls. It has to include people and their cultures" (quoted in Madison, 1993, p. 31). For Richard Moore, cofounder of the SouthWest Organizing Project (SWOP), "our habitat is not a recreational playground. It's where we live and work. It's where our children play and our elders are buried. It's our land and our culture. . . . We care about the outdoors and wildlife, but we care about ourselves and our families, too" (quoted in Beasley, 1991, p. 42, and in Kerr and Lee, 1993, p. 18). Rhiannon Chavis-Legerton, a 13-year-old member of the Center for Community Action of Robeson County, North Carolina, eloquently redefines the environment and human–nature relations:

> My work to protect the environment is justice work, too. The term environmental justice reminds me of the Pledge of Allegiance, when

it says, "and justice for all." To me this should mean the land also, but do people live by that? No. But they are also saying that they love their country. Are not trees part of this country? People forget that we live on their oxygen. Isn't water part of this country? People forget where we get our fish. So to me, the term environmental justice means exactly what it says—justice for the environment. (1993, p. 11)

Environmental justice activists reject the legacy of Descartes and Bacon that separates humanity from nature and designates humanity as the subject that acts upon nature as object; instead, their discourse embraces the gaps in the boundary between humanity and nature and places people in nature as inhabitants (dwellers in place). This discourse is enshrined in the 17 Principles of Environmental Justice adopted in 1991 by the First National People of Color Environmental Leadership Summit. The first principle of environmental justice "affirms the sacredness of Mother Earth, ecological unity and the interdependence of all species" (The Principles, 1993, p. 19). Other principles argue for "a sustainable planet for humans and other living beings," the need to "rebuild our cities and rural areas in balance with nature," and the need to oppose "military occupation, repression and exploitation of lands, peoples and cultures, and other life forms" (The Principles, 1993, p. 19).

The rhetorical efforts of environmental justice activists to dislocate and redefine nature as the places people inhabit and to put humans among other living beings open up unprecedented opportunities for radical critiques of all forms of domination and for grassroots formulations of liberation projects previously constrained by modernism, enabling such groups to focus on issues and deploy tactics that expand what counts as environmental politics. Characterizing mainstream environmental groups as narrow, racist, classist, and bureaucratic, environmental justice activists' redefining of nature/environment authorizes them to care about wilderness and care about themselves and their communities. It enables them to organize on issues that go beyond wildlife and include people. Their goals can include "social and economic justice for all, protecting our environment, natural resources and wildlife habitat" (KFTC activist Daymon Morgan, quoted in Zuercher, 1991, 154). As Dana Alston explained in her address to the First National People of Color Environmental Leadership Summit: "For us, the issues of the environment do not stand alone by themselves. They are not narrowly defined. Our vision of the environment is woven into an overall framework of social, racial, and economic justice" (quoted in Gottlieb, 1993, p. 5).

The rhetorical efforts of environmental justice activists to deconstruct nature and protect local places also have had important practical effects. Nationally, they have basically halted the waste industry's strategy of opening new hazardous waste landfills. CCHW's McToxics Campaign, which included mailing used foam containers to owner Joan Kroc and leaving hundreds on restaurant countertops, led McDonald's "to eliminate its trademark clamshell package, a polystyrene foam container produced with the use of chlorofluorocarbons [CFCs], an ozone-destroying compound" (Gottlieb, 1993, p. 162). They were instrumental in getting right-to-know and emergency planning provisions added to the Superfund Reauthorization Act (Gottlieb, 1993, pp. 190-191).[11] Throughout the United States the kaleidoscopic network of environmental justice groups has worked to protect thousands of communities from potential or existing landfills and hazardous waste sites, as well as other problems (Szasz, 1995, pp. 72–76; Dowie, 1995, pp. 131–135).

A favorite metaphor of the environmental justice movement for describing its goal is "plugging the toilet" of the industrial system, a result that would transform industrial society. Mainstream environmental groups have accepted the parameters of the industrial system and just work to mitigate its most adverse effects, but environmental justice groups question the premises and challenge the practices of industrialism that produce the wastes in the first place. This is a crucial difference, for, as ecologist Barry Commoner observed, "Environmental pollution is an almost incurable disease, but it can be prevented" (quoted in Dowie, 1995, p. 39). For example, while mainstream groups have debated the merits of landfills versus incinerators and the National Wildlife Federation and the Audubon Society invited onto their boards of directors the chair and president of Waste Management Inc. (Gottlieb, 1993, p. 168; Dowie, 1995, p. 219), environmental justice groups have confronted Waste Management Inc. in communities across the country and rejected both landfills and incinerators as "safe" solutions. Instead, they have focused on the practices that produce hazardous wastes, thus transforming the grounds of the debate from technical reforms to a radical critique of the industrial system.

The tactics of environmental justice groups also have started to transform the grounds of environmental politics. In particular, their use of confrontational tactics and local activists have challenged the mainstream environmental groups' reliance on compromise and experts while their focus on environmental justice issues has expanded the possibilities of environmental politics.[12] While mainstream groups advocate economic solutions that grant corporations pollution credits, environmental justice groups hope "their plug the toilet" goal so increases disposal costs that corporations will be forced to change production practices. While mainstream groups practice public relations to polish moderate images, environmental justice groups perform rude and crude tactics to tarnish corporate images. There is some evidence that corporations are vulnerable on both the economic and the image fronts. By the 1990s, waste generators and the disposal industry understood local siting opposition as their biggest problem (Szasz, 1995, pp. 72–73, 103–105). Browning-Ferris decided to get out of the hazardous waste business because it was losing money and damaging its image. McDonald's gave up clamshell packaging to protect its image (Greider, 1992, pp. 167–174).[13]

Finally, while mainstream groups use the public as a direct-mail target to fund their scientific, legal, and lobbying experts, the environmental justice movement *is* local people activated by personal experiences and love of particular places. In collusion with industry and government, mainstream environmental groups have abrogated the legal and political rights of citizens. In 1989, "the Sierra Club helped Senator Mark Hatfield (R-OR) draft an amendment denying private citizens the right to sue against timber sales on the Oregon coast" (Dowie, 1995, p. 214). In North Carolina, the Environmental Defense Fund worked with the state and the biotech industry to draft a law that included an "exclusion principle" that effectively prevented public participation in the regulation of biotechnology (Dowie, 1995, p. 137). Instead of trading away the rights of citizens and communities, the environmental justice movement depends on local citizen participation as it confronts industrial waste practices community by community and advocates the right to a healthy environment. William Ruckelshaus, the first EPA administrator and current CEO of Browning-Ferris concludes, "They are the most radicalized group I've seen since Vietnam. They've been empowered by their own demands. . . . I think what is happening is that people are taking back the power to govern. It's not just symbolic power, it's real power" (quoted in Greider, 1992, pp. 168–169). This move to radical democracy has spawned what sociologist Szasz describes as "*radical environmental populism.* The movement has brought a whole new mass base of working people and people of color to environmentalism. It forged practical and conceptual links between environmentalism and the struggles against racism and sexism. . . . It even envisions a future in which grass-roots environmentalism spearheads the reconstitution of a broad social justice movement" (1995, pp. 6, 166, emphasis in original).

ARTICULATION AND ENVIRONMENTAL POLITICS

To argue that postmodernism makes the establishment of foundations political and recodes neglected practices as political (micropolitics) is not enough to answer the criticisms of sympathetic critics like Harvey

(1989, pp. 46–47) or Jameson (1991, p. 319). They would grant those points about postmodern politics, but for them (and many other critics), "the crucial issue in the politics of the postmodern" is the inability to coordinate local and global struggles (Jameson, 1991, p. 330). Laclau and Mouffe's theory of articulation offers a possible answer to criticisms that postmodernism is a reactionary, aesthetic stance that enervates oppositional politics. Articulation is a way of understanding how, in a postmodern world with neither guarantees nor a great soul of revolt, diverse groups practicing an array of micropolitics can forge links that transform their local struggles into a broad-based challenge to the existing industrial system.

As already discussed, for Laclau and Mouffe, the task of the new social movements involves "expanding the chains of equivalents between the different struggles against oppression" (1985, p. 176). In other words, the new social movements need to disavow an essentialist identity politics that balkanizes and instead link the different antagonisms that give rise to environmental struggles, workers' struggles, feminist struggles, and anti-racist struggles so as to make possible the disarticulation of the hegemonic discourse that constructs these various groups in relations of oppression.

Environmental justice groups have been working to establish contingent alliances directed toward political interventions in larger-than-local discourses. Their redefinition of environment has enabled them to forge links with groups concerned with race, class, and rural issues. As Richard Moore explains, "We perceive environmental issues as racial and social issues" (quoted in Beasley, 1991, p. 42). The articulation of links among different struggles is evident in the actions of civil rights groups and leaders confronting environmental racism. It is evident in the efforts of CCHW's network of over 7,000 grassroots groups "of diverse people and cultures" ("Movement on the Move," 1993, p. 3) challenging the industrial system so that toxics are not in anyone's backyard [a move from NIMBY, "Not in my backyard," to NIABY, "Not in anyone's backyard" [Gibbs, 1993, p. 2]). It is evident in the rhetoric of activists calling for building bridges and finding common ground, for working "to link the struggles so they are born into a movement. . . . We need to bring the focus of all these local struggles together into a regional effort for social change" (Connie Tucker, quoted in Madison, 1993, p. 32).

Environmental justice groups, then, recognize their common struggle against the discourse of industrialism, while simultaneously engaging in local struggles situated in place. They recognize that, as Robert Gottlieb puts it in *Forcing the Spring*,

> the connections among the workplace health and safety movement, the labor movement, and the environmental movement are made in

the industrial choices and practices out of which contemporary environmental problems arise. . . . And while the agendas, organizational forms, and political biases of environmental groups can differ significantly, they still share a common search for a response to the dominant urban and industrial order. (1993, pp. 10, 11)

In this recognition, in their rhetorical practices of constructing nature in ways that lead to linkages and networks among disparate groups, environmental justice groups embody a possible politics in a postmodern social field marked by fragmentation, simulation, and diversity.

My characterization of postmodernism grants that, in a certain sense, Harvey and Jameson are right. Postmodern theorists, in questioning the subject, identities, and grand narratives, transform the grounds and possibilities of politics. Postmodern critiques problematize notions of the revolutionary subject and the privileged class as well as throw doubt upon the possibility of a popular front. However, postmodern critiques also open up politics. What is taken for granted in and founds modern politics—the subject, essential identities, reason, progress, nature—become sites of contestation and the beginning of politics.

Thus, the conditions of postmodernism make possible the deconstruction of modern nature and the articulation of different natures. Following Butler, I would argue for a nature that is not a transcendent foundation, but rather an ideograph that becomes a site of permanent openness and possible resignifications (1992, p. 16), an arena of hope and contestation where people construct and deploy natures for alternative political aims. I believe environmental justice groups are doing this. Their success, however, is not guaranteed. An alternative interpretation of postmodernism as the completion of the modern project would authorize a transcendent humanity to continue progress indefinitely and to consume an obliterated nature as simulations for entertainment. This is the risk and challenge in the wake of modern nature, a risk that calls for political and social struggle on many fronts. If one of the crucial fronts is the mass-mediated televisual public sphere, a pressing concern must be the effectivity of tactical image events in enemy territory. In other words, if GE (NBC) is televising the resistance, how radical can it be?

MEDITATION III

It is necessary that I make (practically, effectively, performatively),
but for you, my sweet love, a demonstration that a letter can always—and therefore ought—not ever arrive at its destination. . . .
It is good that this is the case, it is not a misfortune, it is life.

—Jacques Derrida, The Post Card

CHAPTER 5

PARTICIPATORY DEMOCRACY
IN ENEMY TERRITORY

In modern industrial society people "directly know only tiny regions of social life" and so "people are pressed to rely on mass media for bearings in an obscure and shifting world;" thus, of "all the institutions of daily life, the media specialize in orchestrating everyday consciousness. . . . They name the world's parts, they certify reality as reality" (Gitlin, 1980, pp. 1–2). Given this situation, radical environmental groups would seem to be forced to attempt to appear in the media; however, the mass media do not constitute a liberal pluralist arena for participatory democracy. As participants in and promoters of a dominant ideology perspective, the media are an ideological state apparatus designed to produce a citizenry that accepts the existing economic and social power relations. In such enemy territory, the possibilities of environmental image events would seem to be severely curtailed. Although such a perspective has a certain explanatory power, it is not the whole story. It tends to ignore how the short-term needs of media companies to be competitive and attract audiences may conflict with the long-term goals of the media as an industry or that of capital industrialism in general. Further, different media companies and media forms often have competing interests and goals. Especially significant for this analysis, dominant ideology critique has trouble accounting for resistance and social change. In this chapter I will explore the effects of media ownership and profit orientation, media frames, and media routines and conventions upon environmental issues and groups through close readings of television newscasts on Greenpeace and Earth First!.

MEDIA FILTERS: OWNERSHIP, PROFIT, FRAMES, AND ROUTINES

The last two decades have seen a growing concentration in media ownership (Bagdikian, 1987; Herman and Chomsky, 1988; Manoff and Schudson, 1986; Miller, 1996, 1998; Parenti, 1993). Now, in the 1990s, "eight corporations control the three major television networks (CBS, NBC, ABC), some 40 subsidiary television stations, over 200 cable TV systems, over 60 radio stations, 59 magazines including *Time* and *Newsweek,* chains of newspapers including *The New York Times, The Wall Street Journal, The Los Angeles Times,* and *The Washington Post,* 41 book publishers" (Parenti, 1993, p. 26). Walt Disney Co.'s acquisition of Capital Cities/ABC, Westinghouse's purchase of CBS, and Time Warner's purchase of Turner Broadcasting have further intensified the process. This concentration of ownership is dangerous in principle because the mass media are central to a democracy and an open marketplace of ideas for a diversity of voices. In any field, monopoly reduces the participants. It is even more dangerous in practice when these media owners are multinational corporations driven by profits, most of which are derived from the advertising dollars of other multinational corporations. In such a forum, the voices of those opposed to the vested interests of media corporations and their clients are likely not to be heard. Examples of direct censorship abound. Media mogul Rupert Murdoch, whose media empire reaches three-quarters of the world's population, claims, "The buck stops on my desk. My editors have input, but I make final decisions" (quoted in Parenti, 1993, p. 33). On a large scale, Murdoch's words have been borne out by his actions as he has altered various products of his media enterprises in order not to offend China. For instance, when the Chinese rulers complained about the BBC World Service, Murdoch removed it from his Star TV. Another story illustrates the pervasiveness and process of corporate influence. When two reporters at Fox's WTVT in Florida uncovered surreptitious use of Monsanto's controversial artificial growth hormone in nearly all of Florida's dairy cows, a Monsanto lawyer pressured Fox. As one of the reporters recounts, "Soon Fox lawyers were rewriting our story, and we got a new G.M. and news director. 'We paid $3 billion for these television stations,' we were told by David Boylan, the manager Fox moved in. 'We'll decide what the news is. The news is what we say it is'" (Wilson, 1998, p. 20).

Censorship, of course, is not limited to Murdoch media. Former *Los Angeles Times* publisher Otis Chandler says, "I set policy and I'm not going to surround myself with people who disagree with me" (quoted in Parenti, 1993, p. 34). As former CBS president Frank Stanton admits, "Since we are advertiser-supported we must take into account the general objective and desires of advertisers as a whole" (quoted in Parenti, 1993, p. 35). Chrysler's advertising agency circulated a letter to magazines requiring them to submit articles for screening for possible offensive content to Chrysler (Glaser, 1997, p. 7). Shortly after General Electric bought NBC, NBC aired the documentary "Nuclear Power—In France It Works." NBC's *Today* show did a segment on boycotts, but refused to mention Infact's boycott of GE. As the producer admitted, "We can't do that one. Well, we could do that one, but we won't" (quoted in Cohen, 1998, p. 7).

A more subtle and pervasive influence on the mass-mediated reality of society is media frames, that is, "persistent patterns of cognition, interpretation, and presentation, of selection, emphasis, and exclusion, by which symbol-handlers routinely organize discourse" (Gitlin, 1980, p. 7). Radical environmental groups are constructed and condemned by the media through two frames. The first is the standard media frame of social movements and protesters as "disturbers of order" or "deviants" (Gitlin, 1980; Lewis, 1981/1982; Parenti, 1993). Techniques employed to frame protesters in this manner include labeling, ignoring and undercounting, favoring counterdemonstrators, scanting of content, and trivialization and marginalization (Parenti, 1993, pp. 107-111). The frame of activists as disturbers of the established order so shapes and limits the nature of public discourse that even activists who are victims of terroristic activities (death threats, physical violence) are labeled terrorists. In an example I touched upon earlier, *Time* labeled Andy Kerr, an environmental activist who had received death threats and been burned in effigy, "Terrorist in a White Collar" (Seideman, 1990). In a similar example, Earth First!ers Bari and Cherney, victims of an attempted car bomb assassination, were labeled terrorists in the headline "Earth First! Terrorist Blown Up by Own Bomb" (Rowell, 1996, p. 161). In another example, the title of a feature story on Earth First! in the *Boston Globe Magazine* labels the activists "The environmental guerrillas" (Robbins, 1988, p. 14).

Although all social movements are framed negatively because the media corporations (and government) fear "all mass movements *as such*, because mass movements threaten to get out of control and disrupt the rationales of their power and privilege" (Gitlin, 1980, p. 77), radical environmental groups face a second frame because of their fundamental challenge to the core principles of modern industrial capitalism. The media understand environmental issues, groups, and disasters, indeed, the world itself, through the discourse of industrialism. This frame allows the media to perform a hegemonic role whereby they acknowledge contradictions and challenges to the dominant ideology but only within the parameters prescribed by the dominant ideology so that the dominant ideology survives roughly intact. As environmental columnist Edward Flatteau explains, "There are exceptions, but publishers are basically hostile to environmental protection. It's a threat to their business. Their economic lifeblood comes from advertising revenues, and that means conspicuous consumption" (quoted in Jacobson, 1998, p. 48). This framing process is evident in the ongoing effort to code or frame environmental justice groups as NIMBY ("not in my backyard") groups. This is an attempt to dismiss environmental justice groups as merely an eruption of selfishness and to make struggling against being destroyed by toxic waste dumps, air pollution, pesticides, or nuclear radiation equivalent to opposing low-income housing in one's neighborhood or a sign of pathology. The executive vice-president of the Governmental Refuse Collection and Disposal Association went so far as to deride NIMBYism as "a public health problem of the first order. It is a recurring mental illness that continues to infect the public." He likened environmental justice organizations to "the viruses and bacteria which have, over the centuries, caused epidemics such as the plague, typhoid fever, and polio" and proposed "a campaign to wipe out this disease" (quoted in Dowie, 1995, p. 131, and Szasz, 1995, p. 180).

Syracuse University sociology professor Allan Mazur dismisses environmental justice groups' concerns with nonlocal problems, that is, their move from "not in my backyard" to "not in anyone's backyard" (NIABY), as "'pure rhetoric'—a simple defense against charges of selfishness directed at Nimby groups" (Piller, 1992, p. 115). It is interesting that self-interested individualism ("economic man") is denounced as selfishness when it is not working on behalf of economic interests. Of course, this denunciation is not on behalf of a moral principle of selflessness but rather supports the economic self-interests of large corporations and government officials. So although the corporate-funded Wise Use groups extol the sanctity of "property rights" in cases of development versus environmental protection, when homeowners and local citizens are fighting the excesses of economic development and progress, thus threatening corporate profits, protection of "property" is dismissed as selfish. "'Citizens groups . . . fail also to accept . . . the need for solutions. 'Put it in Texas,' is a convenient argument for local use (unless you're in Texas), but it merely passes the buck and denies the fact that those who benefit from technological advancements must also share the burden of responsible management of its by-products'" (quoted in Szasz, 1995, p. 80).

Contrary to this denunciation, environmental justice groups often quite explicitly transcend local self-interest. Lois Gibbs went from housewife to local activist in the Love Canal incident to founder of the Citizens Clearinghouse for Hazardous Wastes, a national umbrella group for over 7,000 "environmental justice groups . . . a real people's movement" (Gibbs, quoted in Piller, 1992, p. 114). In Arkansas, the Arkansas Chemical Cleanup Alliance, Citizens Against Polluting Streams, Friends United for a Safe Environment, and twelve other such groups banded together to form the Environmental Congress of Arkansas in order to share tactics, resources, and information, articulating their individual fights into a collective struggle "to reclaim Arkansas, a place that the Green Index rates as the third most polluted state in the country" (Webster, 1992, p. 112). There are numerous other examples of cooperation among environmental justice groups that exceed narrow self-interest (see Gottlieb, 1993; Greider, 1992; Kaplan, 1997; Schwab, 1994; Szasz, 1995; and Zuercher, 1991). Indeed, recognizing the global character of corporate industrialism, environmental justice groups "have also begun networking with environmental nongovernmental organizations (NGOs) in the third world. They hope one day to extend their acronym to NOPE (Not On Planet Earth)" (Dowie, 1995, p. 135; see also, Kaplan, 1997). Most important, as I discussed in the previous chapter, the tactic of blocking waste disposal in community after community is designed to "plug the toilet" of the industrial system in order to force a change in production practices. Such a change would be an important step in enacting a just society, wherein a healthy environment would be a basic right for all.

Finally, certain media routines and conventions adversely affect coverage of environmental groups and issues. The supreme journalistic value/method of objectivity, which in practice means quoting official sources, silences or slights environmentalists. A Fairness and Accuracy in Reporting survey of over 800 articles in the mainstream U.S. press found that of attributed quotes, more than 50% were from government officials, the second largest percentage were from industry, and the lowest number, less than 4%, came from environmental groups (McDonald, 1993, p. 7).

The belief in objectivity causes further problems. As Glasser and Ettema (1989, 1993) convincingly demonstrate, the strictures of objectivity prevent journalists from making judgments about values and the moral order. Instead, values must be objectified, transformed into empirical claims that can be reported as facts. There are two important problems with this process. First, values and morals are objectified through an appeal "to some self-evidently credible moral authority" (Glasser and Ettema, 1989, p. 10)—that is, the law, experts, government officials, statistics, common decency—all of which tend to be pillars of the status quo. Activists agitating for social change rarely qualify as "self-evidently credible moral authorities." Second, making mention of values appear under the guise of facts forces underground the putative raison d'être of journalism: the discussion of values and the public good

(Peters and Cmiel, 1991; Carey, 1986; Romano, 1986; Glasser and Ettema, 1993; Parenti, 1993).

The mass media's focus on discrete, spectacular events prevents sustained coverage of the more serious environmental problems, such as global warming or ozone depletion. As media critic Mark Hertsgaard put it, "How do you take a picture of the earth getting hotter?" (1990, p. 18). ABC reporter Barry Serafin concurs: "How do you show soil or groundwater contamination? . . . Sometimes there is no solution, and good stories go unreported because it is too difficult, time-consuming, or expensive to illustrate them" (quoted in Helvarg, 1994a, p. 275). Tom Winship, former editor of *The Boston Globe,* notes: "There isn't a 'Stop the presses!' kind of development on the environment story every day. This is not event coverage. We need to persuade the media to cover the environmental story consistently. Sure, it's a slow story, but they've got to change their attitudes about what makes a story" (quoted in Hertsgaard, 1990, pp. 16–17). This "convention of *describing* an exceptional event rather than *explaining* its sources in normal, everyday social life" (Gitlin, 1980, p. 185) allows radical environmental groups to get attention through image events but it also places the focus on the image events themselves instead of the underlying issues that prompted the groups to protest in the first place. The focus on discrete events decontextualizes the events, further obscuring the connection of events and issues to the underlying systemic practices that need to be changed. As Gitlin points out, "The dissociation of action from context is a central and continuing feature of modern capitalist society. For a movement that intends to transform society, not to enter into its stabilizing rituals, such a dissociation is murderous" (1980, p. 238).

Finally, the news media's emphasis on the new, its quest for the novel, forces groups to perform ever more outrageous image events in order to get coverage. The need for the spotlight of the mass media in order to have a voice in the mediated public sphere may subtly change the focus of a group, so that a major goal becomes getting airtime as opposed to working on local organizing. In other words, "the spectacle always threatens to engulf opposition as soon as opposition turns toward the media for amplification" (Gitlin, 1980, p. 161).

The foregoing analysis reads radical environmental groups as inhabiting a difficult place. Image events are a necessary tactic for oppositional politics in an electronic public sphere. Far from being desperate stunts, then, image events are appropriate actions that have been successful to a certain extent. Because they must appear in a context controlled by corporations, however, it would seem that their radical possibilities are severely curtailed by the constraints of the conditions in which they must operate. As Gitlin puts it, "Just as people as workers have no voice in what they make, how they make it, or how the product is distributed and used, so do people as producers of meaning have no voice in what the media make of what they say or do, or in the context within which the media frame their activity" (1980, p. 3). Most progressive critiques concur with this conclusion.[1] Such a view is not conclusive, however, because it rests on three assumptions. First, it too easily adopts a dominant ideology frame that is too simple to account for a discursive field of competing discourses, what Bahktin calls "dialogic heteroglossia." Second, it assumes transparent meanings and a passive audience, ignoring important audience reception research, especially with regard to television images, that reconceptualize the audience and the process of sense-making. These two assumptions are rooted in an implicit adoption of the discipline's hegemonic transmission model of communication, yet dissemination is more appropriate for a broadcasting model. The first assumption I will address presently. Meanings, audiences, and communication models will be considered in the succeeding chapter.

DOMINANT IDEOLOGY, HEGEMONY, AND FRAMING ENVIRONMENTALISTS: TWO EXAMPLES

Analyses of the political economy of the news media (Bagdikian, 1987; Garnham, 1979, 1995; Herman and Chomsky, 1988; Jhally, 1989) would seem necessarily to lead to a dominant ideology thesis, most memorably put forth by Marx and Engels:

> The ideas of the ruling class are in every epoch the ruling ideas, i.e., the class which is the ruling *material* force of society, is at the same time its ruling *intellectual* force. The class which has the means of material production at its disposal, has control at the same time over the means of mental production, so that thereby, generally speaking, the ideas of those who lack the means of mental production are subject to it. (1956, p. 64)

Gramsci's notion of hegemony provides a theoretical tool for moving beyond the determinism of the dominant ideology thesis and the base–superstructure model. Unfortunately, due in part to Gramsci's ambiguous original formulations, the radical possibilities of hegemony are often undeveloped in favor of readings that render hegemony synonymous with dominant ideology (Gitlin, 1987; Condit, 1994). This mistake is even made by sophisticated social critics, so that hegemony becomes some "thing" a particular group possesses, wins, or imposes, rather than a relational process (this is akin to thinking of power as a thing someone possesses rather than a relation). Stuart Hall, for example, defines hegemony as "that state of 'total social authority' which, at certain specific junctures, a specific class alliance wins, by a combination of 'coercion' and 'consent,' over the whole social formation, and its dominated classes" (1980b, pp. 331–332).

Even in Gitlin's wonderfully complex account of the relations between Students for a Democratic Society (SDS) and the mass media (maybe the problem starts with the monolithic overtones of this latter phrase), his statement above on the people and the media render "the people" powerless ("no voice") and the corporate media all-powerful. This notion of hegemony as dominant ideology runs throughout Gitlin's analysis of SDS, so that in describing hegemony, Gitlin echoes Marx: "those who rule the dominant institutions secure their power in large measure directly *and indirectly,* by impressing their definitions of the situation upon those they rule and, if not usurping the whole of the ideological space, still significantly limiting what is thought throughout the society" (1980, p. 10; see also pp. 253–254). Such a conception of hegemony informs Gitlin's pessimistic conclusion that all groups working for social change face an inescapable dilemma: either be radical, confront the dominant discourse, and be consigned to marginality, or be reasonable and thus be quickly assimilated by the hegemonic worldview (1980, pp. 290–291). From this perspective, hegemony inevitably irons out all contradictory life experiences, competing discourses, contradictory values, and resistances (Gitlin, 1980, pp. 10–11).

Fortunately, in the bulk of his analysis Gitlin does justice to the complexity of activist–media relations, wherein activists are able tactically to exploit the competing discourses and hegemonic routines of a corporate-owned yet heterogeneous mass-mediated public sphere in order to have their fragmented voices heard and to effect social change.

The more nuanced facets of Gitlin's account are beholden to Raymond Williams's explanation of hegemony, because as more than a complicated version of the dominant ideology thesis, hegemony offers a different way of thinking about social relations, politics (micro), power (Foucauldian), cultural activity, resistance, and social change. Williams (1977b, pp. 108–114) explains hegemony as a conflictual process of everyday, lived practices that constitute, renew, and alter a culture's shared reality, its common sense. With this take Williams has highlighted Gramsci's more radical insights, which reconceptualize power as relational, expand the sphere of politics to the everyday, and revalue cultural practices as co-constitutive of the social instead of mere epiphenomena.

Laclau and Mouffe extend these insights. Understanding hegemony as characteristic of politics in change-wracked modern times (1985, p. 138; see also Williams, 1977b, p. 110), they argue that there is no essential "class" that wins or imposes hegemony, but rather that the very identity of classes is formed and transformed through hegemonic practices (1985, p. 58). Similarly, there is no hegemonic center, but, rather, "in a given social formation there can be a variety of hegemonic nodal points" (1985, p. 139). Indeed, a hegemonic discourse can only exist in confrontation with antagonistic articulatory practices, that is, amid competing discourses (1985, p. 135). Such an understanding of hegemony envisions a social field not reduced to a dominant class or class alliance oppressing the subordinate through coercion and consent, but, instead, Bahktin's dialogic heteroglossia or a conflictual public sphere wherein many competing discourses are at play, opening spaces for resistance and change. We can see this in CBS, ABC, and NBC's introductory framing of Greenpeace's encounter with the Russian whalers.

Voice	Image
ABC Anchor: Russia and Japan are the only two nations still engaged in large-scale whaling. A Russian whaling fleet in the Pacific was recently routed, chased away by a fishing vessel manned by ecology-minded Canadians. Bob Marsden of KGOTV in San Francisco tells about it:	Extended medium shot of anchor. In the background a square graphic diagonally split from bottom left to upper right. On the diagonal is the word WHALING. In the upper-left hand corner is the maple leaf of the Canadian flag. In the bottom right-hand corner is the hammer and sickle of the Soviet Union.
CBS Anchor Roger Mudd: A Canadian ecology organization claims a boat it operates chased a nine-ship Russian whaling fleet for 250 miles off the California coast this weekend in order to save the whales. Richard Threlkeld reports:	Extended medium shot of Mudd. In the background a map of the California coast with a dotted line tracing the journey of the Greenpeace ship.
NBC Anchor John Chancellor: Earlier this month an international conference was held on whales and the countries that still hunt whales, mainly Japan and the Soviet Union, agreed to catch fewer of them. A lot of ecologists are worried about	Extended medium shot of Chancellor. In the background a black-and-white sketch of a sperm whale surfacing.

whales, and one group has gone to
war with the Russians. Thirteen
people in an 80-foot boat have taken
on a whole Russian whaling fleet in
the Pacific Ocean. Here's more on
that from Ray Cullen in San
Francisco:

Since I have read the Greenpeace confrontation with the whalers as
a critique of the discourse of industrialism (see Chapter 3), from a dominant ideology or simplified hegemonic perspective we would predict
that the Greenpeace image event would be framed negatively. However,
the historical context of 1975 provides other compelling discursive
frames through which to read the image event. In this case, all three
news anchors frame the Greenpeace action within the discourse of the
Cold War, of democracy versus communism, the free world versus the
evil empire. Instead of being condemned as troublemakers for disturbing the industrial order, the Greenpeacers are embraced as heroes, in-
trepid individuals (thirteen people against a Russian fleet) who went to
war against the Soviet Union and returned victorious, routing and chas-
ing away the Russian fleet. An extended analysis of the CBS clip will
show how this heroic Cold War framing is developed.[2]

Image	Voice
1. Long shot of majestic Greenpeace boat gleaming white in the sun as it crosses under the Golden Gate Bridge.	*Threlkeld*: The Greenpeace expedition, which spent some years sailing around the Pacific
2. Medium shot of the side of the ship.	trying to stop American and French nuclear tests, is now busy trying to stop the Russians
3. Close-up of left bow of ship with crew members at the rails. One waves.	from the massive hunting of whales off the California coast.
4. Long rear shot of the ship sailing into the sunrise. City skyline is in the background.	They arrived in San Francisco today
5. Medium shot of the ship's bow.	after a week of playing cat-and-mouse with the Soviet fishing fleet
6. Medium shot of the ship's side.	and from film

7. Zoom in on black-hulled Soviet factory whaling ship.

taken aboard the Greenpeace boat last Friday, this is how it has been going:

8. Close-up of side of Russian ship.

the Soviet whaling ship harvesting its catch

9. Long shot of the Soviet whaling ship from Greenpeace vessel. Ropes of Greenpeace boat and Canadian flag in the foreground. Zoom in to the metallic gleam of the harpoon gun on the prow of the whaling ship. Whaler mans the harpoon.

with the Greenpeace close by. A Soviet chaser ship armed with harpoons chases a herd of sperm whales.

10. Medium shot of rubber Zodiac with standing crew member.

High-powered rubber rafts from Greenpeace harass the Soviet boat,

11. Long shot of Russian factory ship with Greenpeace Zodiac in foreground.

skimming along like so many water flies between the Russians and the whales.

12. Long sweeping shot left to right from harpoon gun following the flight of the harpoon over the Zodiac and into the whale. Splashes in the water, first from the harpoon and then from the whale.

The Greenpeace managed to save eight whales by using this tactic, but finally the Russian harpooner, whale in his sights, fires, the harpoon flashing only 15 feet above the head of the Greenpeace crew and moments later a dying whale thrashes in the water.

13. Picture of docked Greenpeace boat with crew, media, and onlookers mingling on and around it.

The crew was surprised and frightened at the Russian response.

14. Close-up of Bob Hunter, with a big bushy beard, looking to his right and addressing the reporter off-camera. Imposed on image is "Bob Hunter Greenpeace Project Director."

Bob Hunter: We had really expected that they would not shoot if there were human beings in the way and they did, so we're presuming that when we go out again that perhaps at some level somewhere they would rethink

15. Medium shot panning left to right of Greenpeace crew members on docked ship. Crowd of reporters and onlookers in foreground.

16. Close-up of two crew members, one with wild hair and mustache and the other with a bushy beard and wearing a bandana. Both wearing ragged jackets, possibly army fatigues. After a few seconds, they quickly turn their heads right.

17. Woman on board rushes to rail and leans over to hug man coming to boat. Another man on ship stands in background.

the proposition and when we do it again they might hold back on their harpoons.

Threlkeld: In a few days the Greenpeace expedition plans to sail out for another confrontation with the Soviet fishing fleet, hoping that by doing so it may prompt Washington and other governments to take some kind of action, if not for the whales, then simply to avoid another

international incident with the Russians. Richard Threlkeld, CBS News, San Francisco.

The heroic Cold War frame is immediately affirmed in the opening images of the news report. The first image is a striking shot of the shining-white Greeenpeace vessel steaming triumphantly under the Golden Gate Bridge and into San Francisco Harbor. It is a scene reminiscent of returning World War II veterans, especially with the close-up of the crew in the third shot. In the fourth shot Greenpeace sails into the sunrise, heralding the dawn of a new day. Threlkeld's opening commentary simultaneously casts Greenpeace as Cold War warriors, suggesting that while once Greenpeace may have been concerned with American and French nuclear testing, *now* they are busy trying to stop the Russians from whaling, an activity that positions them alongside the military, political, and industrial institutions of the West that are also "busy trying to stop the Russians."

This frame is reinforced with the typecasting of the Greenpeace members as rugged individuals, a key mythic character of capitalism and democracy, versus the massive, depersonalized, technological juggernaut of Soviet communism. We are presented with repeated images

Armed with a video camera, Greenpeace activists and their Zodiac are overshadowed by a Russian whaler as it attempts to harass the larger boat.

of the medium-sized, white Greenpeace ship in contrast to the gigantic, black-hulled Russian factory ship and the large, black Russian harpoon ship. Both Russian behemoths are shown towering over the tiny rubber Zodiacs. The Greenpeace crew members are personalized. We can make out their faces. We see them smiling, hugging, waving, and inter-acting with media and bystanders. While the crew members look a bit unkempt, people returning from sea tend to look unkempt. Plus, in the wake of the 1960s, the Vietnam War, and Watergate, a countercultural look may not have resonated so negatively. The Greenpeace director is interviewed. Finally, their compassion for and courage in saving the whales gives the crew idealized human qualities.

In contrast, we seldom see the Russian whalers and when we do they are only shot from a distance, which renders them tiny automatons, clad in somber, dark clothes, enslaved to the technology of their ships. Finally, the crucial act, the firing of the harpoon despite the proximity of the Greenpeace activists, objectifies the Russians as inhuman. This is highlighted by the Greenpeace crew's incomprehension in the face of the Russian act.

Greenpeace suffers one fate common to activists attempting to attract media coverage. In turning the complex issue of whale extinctions into an event likely to get media coverage, the focus on the image event eclipses the issue so that aside from the brief snippets "in order to save the whales" and "trying to stop the Russians from the massive hunting of whales," there are no verbal explanations as to why Greenpeace is trying to save the whales, why the whales need to be saved, why Greenpeace has targeted the Russians, which whales are endangered, and what is Greenpeace's critique of industrial civilization. As described earlier, however, the images do function as rhetorical statements containing Greenpeace's critiques of industrialism, anthropocentrism, and progress.

The Cold War frame allows Greenpeace to avoid the strictures of objectivity, what Parenti (1993) terms "false balancing." Within the Cold War frame, the news media do not assume that there is another side that deserves to be heard in the interests of fairness. So in this instance, there is no balancing interview with a Russian whaler. Instead, Greenpeace activist Bob Hunter's interpretation of the incident is uncontested. The Russians are not afforded the opportunity to claim that whales are not endangered, that whales are a natural resource that will go to waste if not harvested, or that whaling is a Russian cultural tradition. We are not shown images of Russian families and told that whaling provides jobs for ordinary folk and Greenpeacers are just elitists costing people their livelihoods. There is no counterexplantion that whales are endangered because of ocean pollution (of which the United States is a major source).

In this notable example, then, Greenpeace escapes the dilemma that Gitlin posits for all groups working for social change: be radical and marginalized or be reasonable and assimilated.[3] Taking advantage of the public sphere as the structured setting for competing heterogeneous discourses, Greenpeace performs and gets favorably aired an image event that is a radical critique of industrialism, nature as a storehouse of resources, and anthropocentrism because it fits within a Cold War discursive frame. Although the Greenpeace message is not assimilated, it is also not fully voiced but rather subsumed under the Cold War frame. Subsumption is not identical to assimilation, however, and the difference enables Greenpeace to achieve two important effects. First, and most significant, Greenpeace's radical critique of industrialism is aired positively. This is no small feat. For performing an act challenging the fundamental assumptions of industrial culture, Green-peace activists are hailed as heroes. Although this imagistic critique of industrialism is framed within Cold War politics, the critique is still present in all its radicalness and, as will be explored in the next chapter, audience research suggests that key images have lasting resonance while verbal frames are of little long-term import. Second, the positive coverage of Greenpeace not only certifies its existence, it lends the group a certain legitimacy and credibility that will protect it from marginalization when it presents critiques of industrialism that impinge on U.S. interests: protests against nuclear testing, nuclear waste, toxic waste, ocean dumping, the slaughter of dolphins in tuna nets, and so on.

Of course, the presence of competing discourses does not prevent the frequently negative framing of the actions of environmental activists. Indeed, considering that corporations own the mass media, thus controlling the means of material and mental production, it is not surprising that radical environmental groups are predominantly negatively framed since their goals are inimical to the economic interests of these corporations. In other words, what a dominant ideology thesis would predict often does occur. So, for example, when Greenpeace performs image events that fall outside a Cold War frame and critique industrial practices generally perceived to be in the best interests of the United States or powerful corporations, they tend to be framed negatively and marginalized through the techniques of the news media previously mentioned.

A more compelling example would be the consistently negative national news coverage of Earth First!, a group whose critiques challenge lumber, mining, drilling, and cattle interests and question the practices of the U.S. Forest Service. An in-depth analysis of an ABC News story will reveal how Earth First! gets framed. In light of the thoroughness of the framing, we will be forced to ask how Earth First! and its causes receive any public support. An answer requires a turn to audience research and consideration of the process of sense making.

On August 10, 1987, ABC's *World News Tonight* with Peter Jennings, introduced Earth First! to the national public with a 4-minute and 20-second Special Assignment titled "War in the Woods." The coverage by ABC News, which had succeeded CBS as the clear television news leader, became the model for later national news coverage (see CBS News, 1990; NBC, 1990a, 1990b; PBS, 1990; ABC News, 1993). For purposes of analysis, I will divide the report into four sections: (1) Introduction; (2) Theater in the Woods; (3) Naming Earth First!; (4) Passing Judgment.

Image	**Voice**
INTRODUCTION	
1. Close-up of Peter Jennings. Office and people in the background as well as a graphic in the upper right corner of a color sketch of a felled giant tree in a forest. Words WAR IN THE WOODS superimposed.	*Peter Jennings*: Now, the War in the Woods. In the last decade or so the environmental movement has grown increasingly more activist and we have seen protesters interfere with whaling ships. We have seen other activists break into animal research laboratories. Tonight we have a report on a group which says it is so angry with destruction in the nation's woodlands and forests that it has been particularly extreme fighting back.
2. Full-screen shot of graphics. Earlier graphic in lower right corner, words moved to the left. In upper half of screen is the logo of the earth with the words ABC News. To the right, the title SPECIAL ASSIGNMENT.	ABC's Ken Kashiwahara is on special assignment.

In Jenning's introduction, Earth First!, as the unnamed environmental group, is framed as dangerously violent. This is done by means of the title, "War in the Woods," and through the description of an escalation of environmental activist tactics from interfering to breaking and entering to unnamed tactics that are "particularly extreme." Since to "break into" labs is a criminal activity, the audience is left to wonder what activity is even more extreme. As in most stories on social activists who are read as disturbers of a taken-for-granted legitimate order, this story on Earth First! has already become a crime story.

Image	**Voice**
THEATER IN THE WOODS	
3. Graphic of cut tree slides right to reveal pristine forest stream sparkling in the sunlight. This scene grows to fill the screen.	*Ken Kashiwahara*: In the wilderness of the American West it has been a summer of guerrilla theater (sound of people howling).

4. Close-up of bearded man in grunge clothing and camouflage cap howling. He is chained to a cable that appears to be part of a large piece of logging machinery.

Protesters chaining

5. Medium shot of man chained to logging equipment. To his right part of another protester can be seen on the machine.

themselves to logging equipment,

6. Long shot of forest zooms in to tree-sitter standing on a narrow platform high up in a tree above the banner SAVE THIS FOREST.

sitting in trees hundreds of years old to prevent them from being cut down.

7. Close-up of tree-sitter perched above the banner (wearing winter wool cap).

Protester. I've seen too much of it disappear and I can't take it anymore.

8. Person in winter coat and hat feeding by hand a bearded, bespectacled man in a blue winter hat who is buried up to his neck in gravel in the road (see picture on page 125). On the pile of gravel he is buried in is a banner reading: N. Kalmiopsis Wilderness no motorized vehicles beyond this point. A campfire is in the background.

Kashiwahara: They have blocked a logging road by burying themselves in gravel,

9. Side close-up of the heads of two protesters buried back-to-back.

two protesters chained together at the neck

10. Close-up of the head of the man in the blue cap talking.

Buried protester. Defending what's left of the wilderness, defending what's left of the world.

11. From aft, slow pan left to right of two brown-uniformed officers handcuffing two protesters. One protester has a long beard. One

Kashiwahara: They have been arrested and jailed

officer is silver-haired. In the
background is a banner, a mound
of dirt, and a Caterpillar bulldozer.

12. Close-up of wrists being
    handcuffed.

13. Two men, two women, a young   and confronted by angry loggers
    girl and a preschool girl jumping   prevented from working,
    out of rocking chairs in the
    middle of the road and trying
    to get out of the way as a white
    pickup truck comes speeding up.

14. Medium rear shot of man with   including one man who lost his
    an artificial arm in jeans, t-shirt, arm in a logging accident.
    and red suspenders walking up
    to the previously mentioned
    group of men, women, children,
    and rocking chairs in the road.

15. Medium shot of man with   *Logger:* You don't want 'em
    artificial arm gesticulating   loggin'! You live in a mobile
    wildly with real arm. He is   home made out of wood, your
    holding a cigarette. White   pay checks are printed on paper
    pickup truck and forest in   pulp, and now you're down here
    the background.   protestin' a man makin' a livin'!

The first eight shots of the actual news report are largely favorable to Earth First!, both visually and verbally. Visually the audience is introduced to three key image events: machine-chaining, tree-sitting, and road-burying. Further, three of the activists are personalized by being shot in close-up while performing these acts. Importantly, two of the activists are allowed to address the audience. Both present themselves as reasonable people forced to resort to these desperate acts in reaction to forces of destruction. They come across not as unruly and irrational but as thoughtful, committed, and resigned. The blue-capped activist buried in the road is particularly effective as he ends his eloquent statement with a sigh and a smile.

These shots implicitly contain Earth First!'s radical critique of industrialism. The protesters put their bodies on the line in solidarity with trees and ecosystems (nature) and in opposition to the technological march of Progress. This embodied and embedded defense of nature be-

In an action akin to those shown in the ABC News broadcast "War in the Woods," Earth First! activists lock themselves to bulldozers to stop the environmentally destructive process of salvage logging in All Species Grove in the ancient Headwaters Forest.

lies anthropocentrism's abstraction of "man" from the natural world and contests science's contextless universalization of nature.[4] The Earth First!ers' passionate embrace of trees and ecosystems defies the dominant assumptions that nature is a storehouse of resources and progress consists in exploiting the storehouse.

The report abruptly adopts a more negative perspective in shots 11–15, where Earth First!ers are shown as lawbreakers and obstructors of economic progress. In shots 11 and 12 two anonymous (literally faceless in the medium shot from behind) Earth First!ers are arrested, with a close-up of the handcuffed wrists emphasizing the restoration of law and order.

Shots 13–15 decisively move the tone of the news story against Earth First! as they deftly introduce the theme of the environment versus jobs by embodying it in the vivid confrontation between the one-armed logger and the six Earth First!ers. In shots 14 and 15 the Earth First!ers are reduced to a muted background to the angry, foregrounded presence of the one-armed logger. The passionate rage of the logger, who literally spits out the words "paper pulp," evokes compassion since it is clear he has paid dearly for his job and to lose it now because of

Protest marches are a common Earth First! rhetorical practice.

tree-hugging elitist environmentalists seems unfair. The logger, with his blue jeans, suspenders, white undershirt, cigarette, and pickup truck is a clear synecdochical representation of workers in general. The logger is allowed to articulate the commonsense position that we are all dependent on wood products, thus we need to log trees. This, of course, ignores the specificity of Earth First!'s protests against the logging of old-growth forests and wilderness areas. These unique and rare ecosystems provide an insignificant portion of the United States' wood supply. The details of Earth First!'s position never get aired.

Image	Voice
Naming earth first!	
16. Wide long shot of approaching group of protesters carrying banners. Some read: "IN WILDNESS IS THE PRESERVATION OF THE WORLD" [Thoreau]; "STOP MAXXAM"; "CLEAR-CUTTING MUST STOP."	Protesters singing "This Land Is Your Land."
17. Close-up of protesters marching by. Unkempt beards, long hair, and ragged clothes of male marchers lends the protest a countercultural or hippie air. More signs can be read: "EARTH FIRST!"; "SAVE VIRGIN REDWOODS."	*Kashiwahara*: (speaking over Earth First!ers singing "This Land Is Your Land") They are members of Earth First!, a radical, loose-knit environmental movement who believe preserving trees and wildlife are essential to insuring the future of man.
18. Medium shot of clean-cut reporter. In background green trees framed by clear blue sky. Words imposed at bottom: Ken Kashiwahara Grants Pass Oregon.	But these forests are more than just a stage for environmental guerrilla theater. They are a battlefield for guerrilla warfare. Earth First! even has a name for it, ecotage, or sabotage in the name of ecology.
19. Zoom in on spike spray-painted green that is embedded in a big tree stump.	The most controversial ecotage tactic is tree-spiking, pounding nails or spikes in trees.
20. Top-to-bottom pan of old-growth Douglas fir. Near the bottom of the trunk is spray-painted the word "NAIL" in neon orange.	It doesn't hurt the trees, but could shatter chain saw blades, forcing
21. Three men in forest in yellow rain gear and hard hats. One is sweeping a downed tree with a metal detector.	logging companies to use metal detectors

22. Close-up of metal detector.

23. Log being milled by a bandsaw.

24. Long shot of log being automatically processed at lumber mill. One worker in hard hat in picture.

25. Zoom in to extreme close-up of mangled saw blade.

26. Young mill worker with bandaged lower jaw. Slow pan left to young woman next to him on couch.

27. Zoom in to close-up of spike in half-processed log.

28. Man in yellow rain coat and hard hat using ax to peel away at standing tree's bark.

29. Blunt head of ax hitting spike.

30. Close-up of warning note: "WARNING ALL THE TIMBER IN THIS AREA HAS BEEN SPIKED."

31. Close-up of book cover: Illustration of two shadowy figures holding a wrench and pliers approaching a grader. Book title: *Ecodefense: A Field Guide to Monkeywrenching*. Book opened to pictures of tree-spiking.

32. Close-up of bearded man. River and forest in background. Words imposed: MIKE ROSELLE EARTH FIRST!

to find the nails or abandon tree-cutting operations.

In California

earlier this year several nails went undetected at this lumber mill,

ripped apart a saw and nearly ripped apart a millworker's face.

George Alexander (injured worker): I could've died from this.
Woman: They're trying to save a tree, you know. It's his life and he's just tryin' to make a living.

Kashiwahara: Earth First! denies responsibility for this tree spiking.

In fact, some members condemn the practice,

but others defend it

as long as loggers and the Forest Service are warned ahead of time.

An Earth First! book even supplies instructions on tree-spiking.

Mike Roselle: I've spiked 800-year-old trees to prevent them from being logged by the Forest Service on public lands.
Kashiwahara: And what is your justification for doing this?
Roselle: Because I think it's a worse crime to cut them down.

33. Close-up of man in tie. Words imposed: WARREN OLNEY U.S. FOREST SERVICE.	*Warren Olney*: To discuss a problem is much better than to go to war over it.
34. Extreme close-up of Olney.	Ultimately somebody could be murdered in this whole event.

In this section of the report (shots 16–34) the nascent negative framing of Earth First! is deepened and reinforced. For the first time Earth First! is named and associated with the rather commonsensical and even bland belief that "preserving trees and wildlife are essential to insuring the future of man" (shot 17). However, the words are juxtaposed with images of a rag-tag group of marchers whose dress and grooming mark and marginalize them as outside the mainstream. Immediately after naming Earth First!, the report's focus shifts from guerrilla theater (image events) to guerrilla warfare or ecotage.

The rest of the section, basically the whole middle third of the report, is obsessed with ecotage, particularly tree-spiking, and the possibility of humans getting hurt. It is worth noting that tree-spiking is one of the few ecotage tactics that can possibly harm people. For example, the popular ecotage activity of adding a gritty substance (often sand) to a machine's oil or gas tank will only harm the machine. The key shot in this section is of a millworker seriously injured when a spiked tree destroyed a mill's saw blade. The worker, with a bandaged jaw, notes that he could have died and the woman next to him suggests he doesn't deserve to die because "he's just tryin' to make a living" and, besides, trees are not of comparable worth to humans (shot 26).

This shot evokes compassion and encapsulates three key arguments. First, it is easy to identify with an injured person, especially when the person seems to have been unfairly injured by a random act of violence. Earth First!, then, is associated with random violence and, by extension, with the deviant, inherently evil "monsters" who stalk the nation committing acts of random violence. Though in the next cut the reporter mentions Earth First!'s denial of responsibility, the denial is effectively undercut in the succeeding cuts, which note that some Earth First! members defend tree-spiking and that an Earth First! book supplies how-to instructions. To top it all off, in shot 32 Earth First! cofounder Mike Roselle proudly admits to the crime of tree-spiking.

In practice, Earth First!'s actions are nonviolent. In this particular incident, the tree-spiker who caused the mill worker's injury was caught. He was a Republican with libertarian tendencies. He had no connections to Earth First!. In over a decade of tree-spiking, no one has been killed and I know of only one injury. This is not particularly surprising since the purpose of tree-spiking is to protect trees, not to hurt loggers. Consequently, tree-spikers spray-paint spiked trees and send warning letters to the U.S. Forest Service, loggers, and the media. However, hundreds of loggers do get hurt and killed each year. This also is not surprising considering the industry's emphasis on profits at the expense of people. For example, the logger injured by the spike, George Alexander, almost did not go to work on that fateful day to protest the lack of response to his complaints about what he considered to be a dangerous saw: cracked, wobbly, and in need of replacement. Alexander explained that the saw hit some sort of metal four times a week and "If it had been a good saw, it would've handled the spike better" (Foreman, 1991a, p. 152).

Shot 26 puts in its starkest terms Earth First!'s ecocentric challenge to industrial culture's anthropocentrism. The woman, in her pleading exasperation, is asking how anyone can value a tree over the life of a human being. How can anyone identify with a tree over a person? Her tone suggests that tree-spiking is an act of such lunacy as to be incomprehensible to ordinary people. Her sentiment is reinforced by the sequence of image events from shot 23 to 26. While the shots 23 and 24

Image of a logger injured by a tree-spiking. Though this incident was used to paint Earth First! as a terrorist organization, the actual perpertrator turned out to be a Republican acting on his own.

showing the slicing up of a log probably do not evoke an emotional response of empathy for the pain of the tree (even the idea of a tree in pain seems alien if not ridiculous), the extreme close-up of the mangled saw blade in shot 25 seems ominous, and the shot of the sliced-up worker (even when covered with bandages) does evoke an emotional, empathetic response, especially as he struggles to speak through his injuries of his brush with death. Of course, corporate capitalists adhere to an Orwellian anthropocentrism, where all humans are of the highest value, but some are more valuable than others. Though humans are more important than trees, not all humans are more important than profits. So while over a decade of tree-spiking has killed no one and injured only one person, decades of logging and milling practices have killed and maimed thousands of workers. For example, in 1988 in the state of Washington there were 28 fatalities. One logger in six will have his or her career ended by a crippling or fatal injury (Foreman, 1991a, pp. 153–154). According to *Accident Facts,* in the Forestry and Forestry Services industries there is on average 1.25 cases per 100 workers of accidents involving days away from work and deaths (National Safety Council, 1995, p. 66). Accidents increased in the 1980s as logging corporations busted unions, thus weakening union efforts on behalf of worker safety (Foreman, 1991a, p. 154). In sum, while successful ecotage should not injure anyone (as Foreman argues, "It's nonviolent, because its directed toward inanimate machines" [quoted in Robbins, 1988, p. 14]), successful logging operations figure in loss of lives and limbs as part of the price of doing business at a profit. These dangers are never mentioned in the broadcast, though the logger's artificial arm provides silent testimony.

Finally, in shot 26 the words of the woman seated beside the worker—"he's just tryin' to make a living"—articulate again the logging industry's position that the issue is jobs versus the environment. (This position is also present in shots 20 and 21, where it is explained that tree-spiking forces logging companies either to go to the expense of finding the nails or to "abandon tree-cutting operations.") Environmental protection costs workers their jobs and in extreme cases, as the image of the injured worker suggests, it costs workers their health—if not their lives. Such an articulation transmogrifies the struggle against industrialism, so that the focus is not on the victims of industrialism—ecosystems, animals, workers, people living in polluted habitats—but rather the focus is on the "victims" of environmental regulations and environmental protest actions: workers and corporations. The explicit vocalization of the jobs argument rebuts the Earth First! position implicitly presented in shot 24, which shows a log being automatically processed while one worker watches. This image illustrates Earth First!'s position (which is

In a typical act of nonviolent civil disobedience, a human chain of Earth First! protesters, their arms linked by cement tubes, block Fisher Road, one of the main entrances to the Headwaters Forest.

never vocalized) that automation and the export of unmilled logs, not environmental protection, are costing workers jobs and devastating logging communities. Yes, workers are victims, but victims of the practices of transnational corporations. Earth First!'s contention is borne out by the fact that during the 1980s more lumber than ever was cut in the Northwest yet the number of logging and mill jobs declined sharply. For example, "between 1979 and 1989 the timber harvest on federal lands in Oregon increased 18.5%, in that same period employment in the wood products industry dropped 15%" (Jeff DeBonis, a timber sale planner, quoted in Foreman, 1991a, p. 132).

The final three shots of this middle section of the report deepen the association of Earth First! with violence. Mike Roselle admits to spiking trees and though he gets to explain why, he also accepts the word "crime" as a descriptor for tree-spiking. In shots 33 and 34 U.S. Forest Service employee Warren Olney suggests discussion is the answer but notes "war" is an alternative and in his ambiguous final sentence warns or threatens that "somebody could be murdered." Olney's remarks can be read as a preemptory justification for a violent backlash against Earth First!.

Image	Voice
PASSING JUDGMENT	
35. Long shot from behind of circle of protesters holding hands in front of a loaded logging truck. Forest in background.	*Kashiwahara*: Earth First! was organized 8 years ago

36. Close-up of circle of protesters. Dog in center.

to speak for the most radical fringes

37. Close-up of woman in circle wearing Earth First! t-shirt with clenched-fist logo. She is framed by the grill of the logging truck.

of the environmental movement

38. Two people, obscured by leaves, hooded sweatshirts, and bandannas, speaking to a microphone.

to support saboteurs called "ecoteurs" who have

39. Medium shot of charred remains of a helicopter. Field and mountain range in background.

burned helicopters and

40. Close-up of helicopter propeller and other charred remains. Airport and mountains in background.

damaged bulldozers to save the wilderness.

41. Close-up of a bearded man (Dave Foreman) howling.

Foreman: (howling).
Kashiwahara: Earth First! founder Dave Foreman believes the laws of nature are higher than

42. Close-up of another bearded man howling.

the laws of man (voice over man howling).

43. Medium rear view of Foreman in jeans, black t-shirt, and cowboy hat addressing an audience.

Dave Foreman: Human beings have no right, no God-given right to totally pave, to

44. Frontal close-up of Foreman speaking. Words imposed: DAVE FOREMAN EARTH FIRST!

totally destroy, to totally use, to totally manipulate every square inch of this planet.

45. Three protesters carrying a banner race to a pile of old-growth logs. Security officers chase and apprehend them as they reach the pile. One woman protester, security officer in pursuit, scales the log pile and jumps jubilantly on the summit. She is apprehended.

Kashiwahara: (wild, high-pitched yelling in background) Earth First! says its acts of civil disobedience are necessary because discussion and negotiation have failed. Mainstream environmentalists say Earth First! is giving the entire movement a bad name.

46. Close-up of a clean-cut, shaven man in a charcoal business suit with red power tie. Stuffed snowy egret in background over man's right shoulder. Words imposed: JAY HAIR NATIONAL WILDLIFE FEDERATION.

Jay Hair: I reject out of hand their being environmentalists. They're terrorists, they're outlaws. They should be treated as such.

47. Close-up of bearded Mike Roselle in a t-shirt. River and forest in background. Words imposed: MIKE ROSELLE EARTH FIRST!

Roselle: I think the real terrorists are the ones who are threatening the viability of this ecosystem and those are the people right now who are logging.

48. Medium shot of bearded protesters in ragged clothing marching down a logging road with the banner: SAVE OUR OLD GROWTH EARTH FIRST!

Kashiwahara: Whether terrorists or freedom fighters, Earth First! says

49. Long rear view of five marchers as they confront a bulldozer moving up the road.

it will continue this summer's campaign of conservation through confrontation,

50. Pan right to left from law enforcement officers (U.S. Forest Service patch and sheriff's badge are visible). to three protesters (two men, one woman) sitting and holding hands in front of the bulldozer.

attempting to save the wilderness and man from himself,

51. Quick shot of three workers, one in suspenders and hardhat, standing on the road bank and looking down on the scene.

vowing to live up to its motto

52. Two bearded men who were blocking the bulldozer are now being led in handcuffs by the law officers up the logging road. One of the protesters appears to be Mike Roselle. Camera follows officer holding on to a handcuffed Roselle as they walk past. Roselle dwarfs the officer.

"No compromise in defense of Mother Earth!" Ken Kashiwahara, ABC News, in southern Oregon.

The final third of the report develops the association of Earth First! with violence introduced in the previous section and then passes judgment on the radical environmental group. Kashiwahara starts the section with a brief recounting of Earth First!'s radical history, with an emphasis on its support of ecoteurs and ecotage. Visually, these shots move from a close-up of a woman wearing a t-shirt with Earth First!'s clenched-fist logo (shot 37) to two comically disguised ecoteurs (shot 38) to the charred remains of a helicopter (shots 39 and 40). The two shots of the burned helicopter are particularly powerful at evoking fear of the violence of Earth First!. The shots are out of context. How was the helicopter destroyed? Why? Was anyone hurt? Whose helicopter was it? As a floating signifier the charred remains are easily linked to images of the destruction of terrorism or war. For me, they evoke the charred helicopter remains in the sands of Iran. Charred helicopter remains symbolize national disgrace at the hands of crazed zealots (people with a different worldview).

From the charred helicopter remains the report cuts to an image of the man morally responsible for ecotage, Earth First! cofounder Dave Foreman. He is howling like a wolf (and inciting others to howl). From Kashiwahara and Foreman himself (shots 41–44), the audience learns that the "leader" of Earth First! rejects both the civil authority ("the laws of nature are higher than the laws of man") and the religious authority ("no God-given right") of Western civilization. In short, shots 38–44 paint Earth First! as beyond the pale of both "civilized" behavior and "civilized" belief. The label for such a group is "terrorist." After an interlude of an Earth First! civil disobedience act characterized by wild yelling and mad dashes (shots 45), Earth First! is named as such.

In shot 46 the report passes judgment on Earth First! as terrorists in a pronouncement enhanced because it is spoken by the leader of a respected environmental organization. With his impeccable grooming, well-tailored business suit, red power tie, and immaculate office, Jay Hair, president of the National Wildlife Federation, embodies the very image of official respectability. Indeed, Hair is the only figure in a suit in the entire report. From his centralized base of power (Washington, D.C.), Hair exiles Earth First! from the community of environmentalists, condemns them as "terrorists" and "outlaws," and decrees that "they should be treated as such." Hair's civilized and violent attempt to silence Earth First! closes off any possibility of dialogue with Earth First! and implicitly endorses violence against Earth First!.

Hair's directive proves prophetic. How are terrorists treated? With repression and force, which is exactly what happens to Earth First!. In the ensuing years it is infiltrated by the FBI, laws are passed making

Jay Hair, president of the National Wildlife Federation, pronounces Earth First!ers to be terrorists and advocates the official, civilized response to such outlaws.

peaceful interference with logging operations in Idaho a felony, and Earth First! activists are arrested, beaten, and bombed.

Responding to Hair's charge, Roselle, in a t-shirt (shot 47), is forced to use the vocabulary of terrorism in his convoluted response. Thus, even while trying to refute the label, Roselle perpetuates the association of Earth First! with terrorists. The reporter also adopts the descriptor "terrorists" (shot 48), and though he offers the alternative of "freedom fighters," he succeeds in ensconcing Earth First! in the vocabulary of war.

More important, the report closes with the judgment that Earth First! is a terrorist organization that warrants a repressive government response (shots 49–52). This is effected through words and images. The segment closes with the reporter noting that Earth First! will continue its tactics of confrontation in defense of wilderness, "vowing to live up to its motto, 'No compromise in defense of Mother Earth!'" (shots 51–52). In the United States' democratic, pluralist culture a refusal to compromise is seen as breaking cultural standards of rationality, reasonableness, and civility. In short, "a sustained refusal to compromise often constitutes a serious rule violation in American culture" (Lange, 1990, p. 474).

Earth First! activists understand their refusal to compromise as the only rational position when over 90% of old-growth forests have already been destroyed. As one Earth First! activist explains, "Destroying the world at the rate we are and poisoning our air and water and land and killing off all the things we depend upon for our own life . . . why should we have to be reasonable in the face of such insanity?" (quoted in Lange, 1990, p. 488).

The report can close, then, with images celebrating Earth First!'s principled, courageous stand against the destructive practices of an insane culture or it can embrace the mainstream valuation of compromise and civility and frame Earth First! as terrorists or outlaws that the society must control in the name of law, order, and civilization. The report chooses the latter option, showing images of a confrontation that ends with the arrest of two Earth First!ers by local police and the U.S. Forest Service. The closing shot is of Mike Roselle, avowed tree-spiker, a.k.a. terrorist, being led away in handcuffs by the forces of law and order. In the end, Earth First!'s refusal to compromise justifies repression.

The preceding close analysis, then, shows how words and images conspire to render the damning judgment that Earth First! is a terrorist group that must be stopped by whatever means necessary. Considering that Earth First! is an avowedly nonviolent group whose major tactics

By transgressing anthropocentric laws in support of ecocentric values, Earth First!ers often risk arrest.

are image events and civil disobedience, this is an amazing construction and judgment. It is also a construction and judgment that presages a violent backlash against Earth First! that results in death threats, beatings, a bombing, and covert U.S. government action.

There are also more general news practices that work against a positive framing of Earth First!. The tendencies of escalation and sensationalization can help account for the shift in the broadcast from a focus on image events and civil disobedience to an obsession with ecotage. Ecotage is not only potentially more dangerous, it is also a relatively new tactic for the new social movements of the mass media age. Ecotage also supports the news media's inclination to treat social protests as crime stories. The broadcast functions in what Gitlin terms "event time" (1980, p. 234) so that even in the favorable opening coverage of image events, the audience is presented with a random string of events lacking any context—chronological or otherwise. The specifics of when, where, and why these image events occurred are never given. The only context is constructed by ABC News: un-American radicals are terrorizing the West. Devoid of context, Earth First!'s tactics and critique of industrialism become just another example of the craziness and random violence littering the American landscape.

CHAPTER 6

AUDIENCES, DISSEMINATION, AND CONTEXTS

Rereading "War in the Woods"

Although the preceding analysis suggests that radical environmental groups inhabit a difficult space within the corporate-controlled, mass-mediated public sphere, it is not a hermetically sealed space. There are cracks and openings for resistance, alternative readings, aberrant sense-making. The analysis in the previous chapter of the Greenpeace–Russian whalers confrontation suggested that one opening is the competing discourses of the heteroglossic public sphere. Work by British Cultural Studies theorists suggests, further, that the discourses and experiences an audience brings to a news report open spaces for them to make resistive, alternative, negotiated, and aberrant readings that challenge the hegemonic framing (Fiske, 1986, 1989; Hall, 1980a, 1994; Lewis, 1985, 1991, 1994; Jhally and Lewis, 1992; Morley, 1980a, 1980b, 1981; Williams, 1977b). At this point, I want to argue that audience research on how people watch and make sense of television suggests that not only do competing discourses weaken the strength of preferred readings but also that the very process of making sense of television news, especially the centrality of images, casts doubt on the possibility of an effective hegemonic frame for a particular news report.

In studying media that employ a mix of words and images, critics in rhetoric and cultural studies have tended to emphasize words and narrative form. A famous example is Hall's discussion (1973) of how words anchor the meanings of news photographs (see also Aden, 1994;

Altman, 1987; Gitlin, 1980; Olson and Goodnight, 1994; Seiter, 1987). Gitlin goes so far as to deride television's reputation as a visual medium and instead concludes that television news is typically an "illustrated lecture" controlled by the verbal narrative (1980, pp. 264–265). This is an egregious error, especially with respect to the study of television, which is an imagistic discourse driven by associative logic or what Barthes terms "myth" (1972; see also Postman, 1985; Brummett, 1994; Szasz, 1995).

For decades, quantitative media research, whatever its weaknesses, has pointed to the dominance of images over words, the visual over the verbal.[1] Recently, Kathleen Jamieson and Justin Lewis, working out of the traditions of rhetoric and cultural studies, respectively, have reconfirmed the primacy of images in televisual discourse. Their audience research studies are telling because both scholars, working out of traditions that value the word, were surprised by the power of images in the sense-making process of audiences.

Jamieson, in her ongoing study, is trying to figure out ways to help counteract deceptive political commercials. In one experiment she showed people the political advertisements accompanied by a voiceover critiquing the ads. Jamieson found, to her surprise and dismay, that her method reinforced instead of counteracted the messages of the advertisements. As Jamieson recounts:

> A participant in a focus group had alerted me to this when what she remembered of a network newscast analyzing a political advertisement was not what the reporter debunking the ad had said. Rather, she remembered the ad itself. . . . Subsequent testing confirmed that when an ad was shown on the full television screen while a voice-over pointed out the distortions, viewers remembered the ads better than the corrections. (1994, p. A56)

Lewis's study is even more revealing for our purposes. Lewis had 50 respondents watch an evening news broadcast and then conducted one-on-one qualitative interviews characterized by an open-ended, conversational style. Lewis's conclusions shed light on why hegemonic framing of the news often fails to lead to closure, why, in other words, the broadcasting of Earth First!'s image events is politically potent in spite of the framing. First, people forget almost everything they watch on the news (1991, pp. 124–125; see also Dahlgren, 1985; Szasz, 1995). This is due in part to haphazard viewing habits. As Andrew Szasz summarily explains,

> Millions say they follow the news, but research finds that they do so in a superficial and evanescent manner. . . . Addicted to the consumption of superficial imagery, habituated to a state of distraction, deaf

to complexity and subtlety, the news consumer watches, hears, or reads news stories in a way that preserves, even enhances, their iconic quality: the strong visual and emotional components dominate; attitude formation takes place without much need for detail in the cognitive component. (1995, pp. 61, 63)

Although his description is apt, Szasz's tone is unnecessarily patronizing. Benjamin fruitfully suggests that distraction be considered a mode of perception: "Reception in a state of distraction, which is increasing noticeably in all fields of art and is symptomatic of profound changes in apperception, finds in the film its true means of exercise" (1968b, p. 240). Instead of being condemned as the negative of concentration, distraction is an appropriate form of attention in a culture operating at the speed of technology and immersed in fleeting images. Lewis suggests that this habit of distraction is further encouraged by the lack of classical narrative structure in television news.

Narrative is a central human way of making sense of the world (Barthes, 1974; Fisher, 1987). Narrative is characterized by the code of sequence or logic of development and the hermeneutic code, wherein a question or enigma is posed, sustained, and then finally resolved (Lewis, 1991, pp. 125–126; Barthes, 1974). Ignoring the narrative logic characteristic of most of television, from soap operas to sitcoms, from movies to sporting events, television news adopts the form of newspapers, where the most important elements are presented first. As Lewis observes,

> The hermeneutic code is not only ignored, it is turned inside-out. History inevitably has an enigmatic quality—we do not know how the future will unfold. Television news takes this history and squeezes the sense of mystery right out of it. The main point of the story comes not at the end, but at the beginning. It is like being told the punchline before the joke, or knowing the result before watching the game, or being told "who-dunnit" at the beginning of the murder mystery. (1991, pp. 130–131)

This is often done by the anchor. For example, in the anchor introductions to the Greenpeace news story on ABC and CBS, who, what, when, where, and why are "given away" by the anchor in the opening seconds: Canadian ecologists routed a Russian whaling fleet recently in the Pacific in order to save whales. Deprived of both a chronological sequence and enigma to carry them through the story, how do viewers make sense of the fragmentary, ahistorical bits of "reality" that confront them?

Lewis quickly discounts the notion that the lack of a classical or realist narrative to bind viewers allows them to be free agents in a semiotic democracy. Instead, Lewis finds that three factors constrain viewers as they attempt to make sense of the ahistorical, disparate fragments that constitute the news (1991, pp. 141–151). The first is ideological resonance: "meaning is contingent upon the semiological or ideological resources available to the viewers" (Lewis, 1991, p. 141). Viewers make sense of and remember moments in the news that connect to the larger social discourses that constitute their social environment. Second, echoing the work of Barthes (1972) and Postman (1985), Lewis finds that televisual news discourse is characterized by association. Lacking historical context from either the news itself or larger social discourses, viewers grasp at repeated associations, thus turning the news into myth. Third, the most powerful moment in a news story is what the audience perceives as the event, which is usually the first main action sequence in the report—in other words, the first action images.

Interestingly, years earlier Gitlin had hinted at these insights, but his commitment to verbal discourse blinded him to the potential of images. While arguing that the correspondent situates the story and names the point, Gitlin recognized that the purportedly "decorative and illustrative" images may "testify to the existence of a discordant reality which the correspondent is working to assimilate into a conventional framework" (1980, p. 265). He points to the telling example of the Vietnam War, where images of battles and body bags scorched the verbal hegemonic frame.[2] Wedded to the primacy of the verbal, however, Gitlin concludes that "most of the discrepancies are flattened out by producers and editors. . . . Commonly the lecture is unitary and controlling. The lecture format enables the correspondent and the producer to clamp a rather definitive frame onto a minute or two of film" (1980, p. 265).

Ronald Reagan's media team made no such mistakes in understanding the dynamics of images and words. Indeed, Lewis and Jamieson's insights were already understood and practiced by Reagan's media advisers, especially Michael Deaver.[3] So, while President Reagan was signing a capital gains tax cut, Deaver was scurrying around Boston scouting out the perfect pub for an image of Reagan hoisting a mug of brew with blue-collar workers. As Deaver notes, "The picture that flashed around the world was Reagan hoisting a beer with a bunch of working stiffs. . . . He was a regular guy." This strategy of image politics was used repeatedly by the Reagan media team—images of Reagan at a car race, the opening of a senior citizen center, and the Special Olympics were used to counter the potentially negative public perceptions of Reagan's anti-worker, anti-elderly, and anti-disabled policies and may help explain the dissonance between Reagan's personal popularity and support for his policies. While actively working against the material interests of these groups, Reagan simultaneously courted them with images. From Deaver's perspective, the eye wins over the ear every time.

White House reporter Leslie Stahl, an unwitting pawn in the staging of these image events, says she "just didn't get the enormity of the visual impact over the verbal. No one would hear us," until a White House official congratulated her on a story that contained a biting verbal attack on Reagan accompanied by terrific pictures. The official told Stahl, "They didn't hear you, didn't hear what you said, they only saw those pictures." As Deaver remembers, "She had to put on during her piece all these wonderful pictures we had created. If you really believe that the visual is going to outlast the spoken word in the person's mind, we were delighted with it." Stahl's bosses understood the power of visuals. As Stahl recalls, "It was okay if I said something that was somewhat critical as long as it was covered with pictures, but when I was shown . . . on camera it was to be innocuous, whatever I was saying."

The previous chapter offered a standard rhetorical criticism of the ABC News report on Earth First!. That criticism found that while there were some positive (or at least ambiguous) portrayals of Earth First! in the report, the preferred reading or dominant meaning worked to construct Earth First! as a terrorist organization that must be stopped by the forces of law and order. Further, this construction is typical of other representations of Earth First! in the national public sphere. Yet radical environmental groups and their causes remain popular. In the midst of a double-barrel corporate media atrocity drive[4] on many fronts (television, radio, newspapers, magazines) and a corporate and congressional legal assault in the name of progress and patriotism on environmentalists and environmental protection, most Americans (91%) believe that protecting the environment should be a top or important priority (Public Agenda, 1999), 76% say legislation is necessary to indure that businesses protect the environment (Mitchell, 1996, p. 85), and 67% of Americans agree with the statement "Protection of the environment should be given priority, even at the risk of curbing economic growth" (Gallup/CNN/USA Today Poll, 1999).

Granted, there are many factors that explain this discrepancy. I do not think, however, that we can discount negative representations of radical environmental groups as irrelevant. Instead, we need to explore the weaknesses and epistemological assumptions of a rhetorical approach that treats texts as discrete and assumes (simple models of) intentionality, transmission, and the dominant ideology thesis. In short, a rhetorical reading of news coverage of radical environmental image events cued in to the insights of audience research will not dismiss environmental image events as quixotic assaults on an impervious corporate industrial system, but will instead read such image events as possibly appropriate and effective tactics in a heteroglossic public sphere.

IMAGES, AUDIENCES, AND READINGS

An analysis of ABC's report on Earth First! in light of audience research leads to a radically different reading that points to the potential
rhetorical force of the practice of image events. Following is such a possible reading.

Considering the dominance of images over words, the eye over the
ear, the first step is to focus on images to the near exclusion of words.
This radically alters our earlier reading of the news report, for in that
analysis words were the driving force in the mutation of Earth First!
from a bunch of civil disobedience protesters to a terrorist organization.
Clearly, the reporter is using words to attempt to determine the meaning
of the images. This is similar to Hall's argument regarding the use of
words to determine the meaning of photographs. The distinction I want
to make here is that, yes, news organizations attempt to construct a
hegemonic frame through the strategic use of words to delimit possible
interpretations of images, but this is only a strategy, and one whose effi-
cacy is thrown into doubt by audience research.

Lewis' work suggests not only that we focus on images, but also that
we concentrate on action images:

> The powerful moment in the news discourse is the portrayal of the
> "event"—or at least the part of the story audiences perceive as the
> event. Just as newspaper readers will skim the opening paragraphs for
> the main gist of the news story, so viewers will focus their attention
> upon its televisual equivalent. The equivalent moment, perhaps sur-
> prisingly, does not appear to be the anchor's introduction but the first
> main *action sequence* in the report. (1991, p. 149)

In the example of the ABC News story "War in the Woods," the first
three action sequences (cuts 4–10) are of Earth First! activists performing image events: machine-chaining, tree-sitting, and road-blocking.
The images are largely positive and are clearly of nonviolent civil disobedience. In addition, the voice-over (though of secondary importance) is simply descriptive and two of the protesters are allowed to explain what they are doing and why. In short, through the synecdochical
tactic of image events, Earth First! is able to present itself and its causes
in an extremely favorable light during powerful moments of the news
story.

Also, importantly, the extended segment of the report devoted to
constructing Earth First! as a terrorist organization is bereft of action
images of Earth First!ers performing ecotage. Indeed, in all the action
images throughout the report, the Earth First! protesters are nonviolent

A literally underground Earth First! activist puts his body at risk in defense of what's left of the wilderness.

and twice (shots 11–12, 52) they peacefully submit to being arrested. In fact, the only action sequence that hints at violence is when the one-armed logger (a victim of industrial violence) speeds toward a group of activists blocking a road and then gets out of his pickup and angrily confronts them.

If the audience is using these action sequences to make sense of Earth First!, what sort of sense are they likely to make? Since these action sequences are ahistoricized fragments, meaning will depend on associations made with larger social discourses, on the discourses to which viewers link these fragments. Clearly, the reporter attempts to associate Earth First! with terrorism, but it is an association based on inference and devoid of action images. Still, the proffered reading of Earth First! image events in the "War in the Woods" places them in a context constructed by the discourses of terrorism and law and order. That this is a compelling link and context for the audience is doubtful.

Earth First! simply does not fit these discourses. These are powerful discourses that have been clearly defined. Terrorist acts involve bombings, mass destruction, and shattered bodies. Violations of law and order involve violent crimes: murder, assault, armed robbery, carjacking, kidnapping, and rape. Tree-sitting and burying oneself in a road do not make sense within these discourses. Indeed, the only likely victims are the "perpetrators." Similarly, ecotage, which is never shown, does not resonate with these discourses. It is difficult to equate putting a nail in a tree with blowing up the World Trade Center or a 747.

Of equal significance, Earth First! activists do not fit the most prevalent images of villains in either the discourses of terrorism or of law and order. Conventionally, terrorists are Middle Eastern Muslims. The strength of this identification was evident in the days following the Oklahoma City bombing, when Middle Eastern terrorists were suspected, sketches broadcast, a man of Middle Eastern ethnicity was arrested in the Heathrow airport, and Arab-Americans reported being subject to increased hostility and discrimination throughout the United States.

Similarly, in law and order discourse the monsters to be exterminated are often African-Americans. Their placement in this subject position in the mainstream popular imagination can be seen in the successful Willie Horton political advertisements of the 1988 presidential election and in the assumptions (racial and otherwise) about the "looters" during the Los Angeles uprising (see Fiske, 1994, pp. 169–190). This articulation of black and criminal was drawn out in a *Time* cover story (August 23, 1993), illustrated by a cover that was a caricature of a criminal "monster." Beside the title "America the Violent" is a distorted, snarling figure of indeterminate race: blue, purple, black, and white. The clothes and accessories of this monster, however, mark him as a stereotypically inner-city African-American: hooded sweatshirt, baseball cap, leather jacket, gold necklace, gold rings, and gold bracelets/manacles. The Earth First! protesters, white, apparently middle class, many dressed in tie-dye shirts, Patagonia windbreakers, environmental t-shirts, sandals, and hiking boots, do not match the subject positions of terrorist or criminal as popularly constructed. White, middle-class America[5] looking at images of these activists may not see the Other but their children or themselves in the 1960s.

Another discursive context floated in the report is that of the economic discourse asserting that protecting the environment costs jobs—environment versus jobs. Although this link may have ideological resonance for rural workers who are more likely to be affected by environmental protection efforts, such a discourse is not likely to be meaningful for many urban and suburban viewers. Indeed, urban and suburban viewers are more likely to understand nature not as a source of money and work but as a place to spend money and leisure time. For them, the mainstream environmental discourse that defines nature as an aesthetic and recreation resource may enable them to make sense of the opening image events of the news report. Put in this context, the image of a protester buried in a road saying "Defending what's left of the wilderness" makes sense as a courageous act. The protester is a hero, not a nutcase or someone costing people jobs. He is the defender not merely of wilderness but also of the values of a cultural formation.

The actions and images of Earth First! are easier to link to a discourse of social protest and civil disobedience that gained prominence and respect through the civil rights struggle and the anti-war protests of the 1960s and 1970s. The image events of the early action sequences in the report show Earth First!ers acting in the best tradition of civil disobedience. They are acting peacefully while putting themselves at great personal risk in the cause of interests that transcend narrow self-interests. Further, many of their looks are reminiscent of the student protesters against the Vietnam War.

The tradition of protest points to one final discourse that may affect how viewers make sense of "War in the Woods." Encapsulated discursively on bumper stickers as "Question Authority," the Vietnam War, Watergate, Iran Contra, Waco, the Hill–Thomas hearings, the Packwood diaries, the campaign-funding scandals, the Rodney King beating, the Fuhrman tapes, various big-city police scandals, "downsizing" and corporate flight, NAFTA, and a host of other incidents have helped create a prevalent distrust of authority, of law and order, across the political spectrum that has manifested itself in a range of acts, from not voting and civil disobedience to mailing bombs (the Unabomber) and blowing up the Federal Building in Oklahoma City. A recent *Washington Post*/Kaiser/Harvard survey found that only one in four Americans trust the federal government to do the right thing most of the time, a decline in public trust from 76% in 1964 (Morin and Balz, 1996, pp. 1, 6). From the context of this discourse of distrusting authority, the arrests of Earth First! activists may be read as another example of excessive government repression (often for corporate interests at the expense of "the people," i.e., ordinary citizens). Why is the government arresting a bunch of hippies sitting in the woods (on public lands)?

Which one of these discourses does the audience use? The only thing that can be said with certainty is "not just one." Indeed, probably all these discourses and more come into play as people work to make sense of the image events. My purpose is not to provide the correct alternative reading but to open up the possibilities and provide an example of what it means to say that audiences (we) function in a heteroglossic public sphere composed of competing discourses. (This example also points to the importance of context and history for the critical rhetorician, an issue I will take up in the final chapter.)

Further, I am not suggesting that audience research can give us the truth about what a text means or that rhetoricians need to do audience research. Audience research is useful in that it broadens the scope of analysis. Instead of assuming that meaning inheres in the text, the better audience research understands meaning as a temporary fixing of the negotiations among the text, subjects, and social discourses. Still, like rhetorical criticism, audience research is necessarily a construction and interpretation of those negotiations.

DECONSTRUCTING THE TRANSMISSION MODEL

Rhetoricians need not do audience research but because their criticisms are rhetorical and not merely aesthetic, they ought to perform criticisms in light of audience research. The research of Lewis and Morley, for example, forces a questioning of assumptions about intentionality, meaning, and the audience that points to the beginning of a deconstruction of the fundamental problematic underlying rhetoric and communication studies in general: the transmission model (sender–message–channel–receiver, or speaker–text–audience). Although some may argue that a critique of the transmission model has been done, I think it is still the ground we walk on, which is significant because it is a perspective that hampers our ability to make sense of the rhetorical force of image events. That was evident in my traditional criticism of the ABC News report on Earth First!, in which assumptions about speakers, intentionality, meaning, and the audience guided and constrained my reading. It is evident in Condit's (1994) observation that hegemony has become a synonym for dominant ideology. This is not because people are wedded to the dominant ideology thesis. Indeed, few would defend it in print. Rather, the dominant ideology thesis fits the epistemological and ontological assumptions of the transmission model. In a discipline where our "basic orientation to communication remains grounded, at the deepest roots of our thinking, in the idea of transmission: communication is a process whereby messages are transmitted and distributed in space for the control of distance and people" (Carey, 1989, p. 15), such assumptions exert a certain gravitational force.

This force struck me at a Speech Communication Association panel (New Orleans, 1994) chaired by the then editor of *Communication Theory* (Donald Ellis), when a respected scholar, Robert Craig presented a paper calling for the revitalization of the transmission model. However, the most important instantiation of the dominance of the transmission model is in the textbooks of the discipline.[6] Charles Larson, in the textbook *Persuasion,* notes, "The simplest model of communication, and the one most widely referred to, is the SMCR model suggested by Claude Shannon and Warren Weaver in 1949" (1989, p. 12). In the textbook *Principles and Types of Speech Communication,* the Speech Communication Transaction Model consists of a speaker giving a message through a channel to a listener or listeners who provide feedback (Gronbeck, McKerrow, Ehninger, and Monroe, 1990, p. 13). The best-selling public speaking textbook, *The Art of Public Speaking* (Lucas, 1995), also echoes the SMCR model. Agee, Ault, and Emery define communication as "the act of transmitting information, ideas, and attitudes from one person to another" (1985, p. 18). Berlo's (1960) central textbook is founded on the transmission model. Two later important textbooks on communication, both titled *Human Communication* (Burgoon, Hunsaker, and Dawson, 1994; Tubbs and Moss, 1983), also establish the transmission model as the baseline. John Fiske, in his *Introduction to Communication Studies* (1990), devotes the first two chapters to the transmission model and variations upon it that constitute the central models in the field.

These examples suggest that the transmission model has a certain institutional force. This institutionalization is an instantiation of a long tradition (Peters, 1989, 1996). The extent of this force is most evident when critiques of the transmission model reveal shared metaphysical commitments. For instance, one of the most important lines of critique comes from what can be loosely termed the "medium theory perspective."[7] Medium theorists understand communication and communication media to be fundamentally concerned with constitution rather than transmission and representation. Yet despite this significant difference in orientation, constitutive and transmissive theories of communication share metaphysical commitments that become obvious in the medium theorists' critiques of the transmission model. In *For a Critique of the Political Economy of the Sign* (1981, pp. 179–180), Baudrillard, for example, critiques mass media as transmission systems at a distance and praises posters and notices printed on walls as immediate and thus the "real revolutionary media . . . everything that was an *immediate* inscription, given and returned, spoken and answered, mobile in the same space and time, reciprocal and antagonistic" (1972/1981, p. 176). Baudrillard's emphasis on immediacy, bodily presence, the spoken and answered, idealizes the face-to-face encounter and privileges speech as authentic. This same face-to-face model is the ideal for the transmission model, in which the medium is essentially neutral, transparency is the goal, and feedback mimics the ideal of the give-and-take of the face-to-face encounter. Thus, Baudrillard's critique,[8] in this example, shares with the transmission model a commitment to the dominant metaphor of Western metaphysics, the face-to-face conversation.

What follows is a partial interrogation of the transmission model, not an exhaustive critique. I rely on the previous meditations on image events to question the categories of speaker, intention, text, context, and audience. Derrida's deconstructive project offers a heuristic opening. Derrida suggests that an interrogation of communication, transmission, is "unhearable," a "something that *could not be presented* in the history of philosophy, and which, moreover, is *nowhere present*, since all of this concerns putting into question the major determination of the meaning of Being as *presence*" (1981b, p. 7, emphasis in original). Within that circle of presence is an a priori understanding of communication as transmission, "the vehicle, transport, or site of passage of a *meaning*" (Derrida, 1982, p. 309). Already formed ideas are transmitted between conscious, unified persons: "the exchange of intentions and meanings, the discourse and 'communication of consciousnesses'" (Derrida, 1982, p. 329). The transmission model of communication, with its ontological and epistemological assumptions concerning subjectivity, meaning, and intentionality underwrite communication studies (as well as Western metaphysics) and its areas (mass communication, rhetoric, and interpersonal). The commitment to these assumptions, especially a naïve notion of presence, is most stark in the valorization of face-to-face communication, Dewey's two people on a log, as the ideal baseline form (Peters, 1994). Communication scholars have begun to critique the transmission model of communication and its commitments (Carey, 1989; Peters, 1989, 1994, 1996; Chang, 1985, 1986; Biesecker, 1989b; Angus and Lannamann, 1988; Angus, 1989b). Still, the most sustained critique of communication has been made by Derrida. This critique is most explicit (and "communicable") in "Signature Event Context" (1982) and *Limited Inc.* (1988), although many of Derrida's texts can be read as extended deconstructions of communication (especially *Of Grammatology* [1974/1976], *Dissemination* [1981a], *The Post Card* [1987], and *Writing and Difference* [1978]).

Further, Derrida's work has been characterized by him (1981b, p.

13; 1974/1976, 1982) and others (Poster, 1990; Ulmer, 1989; Spivak, 1976) as a theorizing of telecommunications (electronic media) that rereads the history of media and proposes a general structure for all media and communication. In short, Derrida reads the move from orality to writing to electronic communication as "a more and more powerful historical unfolding of a general writing" (1982, p. 329). The three essential traits of general writing are: (1) a mark iterable beyond the presence of an author; (2) that carries a force of breaking with its context; (3) due to spacing, the disruption of presence in the mark (1982, pp. 317–318). For Derrida, the structure of general writing is characteristic of all texts, written, oral, nonlinguistic. That *there is nothing outside the text* "does mean that every referent, all reality has the structure of a differential trace, and that one cannot refer to this 'real' except in an interpretive experience" (Derrida, 1988, p. 148).

Derrida's proposition is expounded on, as mentioned earlier, by Peters, Poster, and Ulmer. Peters' essay is the most succinct and startling. Proposing to replace the face-to-face conversation as the ideal type of communication, in comparison to which mass communication is always found wanting, with the one-turn, Peters conceives of "mass communication as conversation reduced to its primordial form, the single turn" (n.d., p. 11). From this perspective, "interpersonal communication is simply the splicing together of two or more acts of mass communication . . . two or more people taking turns at broadcasting" (n.d., p. 12). What is primordial for all types of communication is the gap between sending and receiving and, thus, "the radical indeterminacies of reception" (n.d., pp. 12–13),[9] what Derrida understands as every sign's always already breaking of any determinable context due both to iterability and the inhabiting of all signs by *différance*. Communication, then, is characterized not by transmission but by dissemination (Derrida) or broadcasting (Peters).

It is important to remember that Derrida is making an ontological and historical argument about communication.[10] Whether one agrees or not with Derrida's description of the ontological structure of communication does not preclude recognizing the force of his historical argument about the transformation of the character of communication from the presence of face-to-face encounters to the absence of mass media broadcasts. Indeed, for Derrida the ontological structure is only revealed as a consequence of the revolution in historical communication technologies:

> The development of the *practical methods* of information retrieval extends the possibilities of the "message" vastly, to the point where it is no longer the "written" translation of a language, the transporting of

a signified which could remain spoken in its integrity. It goes hand in hand with an extension of phonography and of all the means of conserving the spoken language, of making it function without the presence of the speaking subject. This development, coupled with that of anthropology and of the history of writing, teaches us that phonetic writing, the medium of the great metaphysical, scientific, technical, and economic adventure of the West, is limited in space and time. (1974/1976, p. 10)

Such a historical perspective grants the critic leeway in acknowledging that different models of communication have different valences depending on the historical context.

Given Derrida's historical perspective, his deployment of the term "general writing" is unhelpful due to the connotations it promotes.[11] Derrida's critique of the speech/writing hierarchy aims not at a simple reversal but at a displacement that metaphorically announces the "death of the civilization of the book" as well as the "death of speech" (1974/1976, p. 8) in the face of telecommunications.

I want to think about image events, particularly the example of "The War in the Woods," in the wake of the deconstruction of communication. This is both a theoretical and political intervention. Although Peters notes that "the proper homeland of theory is not the actual but the possible" (n.d., p. 6) (a position denounced by Hikins, 1995, as eristic, within our discipline Derrida tends to be attacked from behind the barriers of common sense (for examples, see Aune, 1983; Ellis, 1991). Ray Birdwhistell, in lamenting the dyadic fallacy, notes,

> Communication has been studied as a process identified by the passage of information through the transmission of more or less meaningful symbols from one individual to another, from one group or representative of a group to another group or representative. Thus, the ideal model for the communicative process is based on the dyad: a knowledgeable monadic father or teacher who emits knowledge-carrying symbols that enter into the head of a less knowledgeable or nonknowledgeable monadic child. Such a conception is deceptively familiar, and it has the absolute support of common sense. (1968, p. 24)

INTENTIONAL FALLACIES AND THE SUBJECT OF RHETORIC

Ellis, in an essay hostile toward poststructuralism in general and Derrida in particular, presents a representative commonsense defense of intentionality and transmission. From such a perspective,

intentionality and communication are inseparable. . . . As soon as intention is considered definitively a part of communication, it becomes impossible to imagine alternative interpretations of a message.
The best way to discover the "meaning" of a message is to turn to the
speaker's or writer's intentions. . . . Communication is by definition
intentional and situated language use, and the importation of a critical theory that ignores these qualities is irrelevant to communication.
(1991, pp. 221–223)

In responding to a similar criticism from Searle 14 years earlier, Derrida
writes, "I must first recall that *at no time* does *Sec* ["Signature, Event,
Context"] invoke the *absence*, pure and simple, of intentionality" (1988,
p. 56). Both Derrida's "Signature Event Context" (1982) and *Limited, Inc.*
(1988, esp. pp.55–79) are *about* "intentional and situated language use."
As Derrida explains in "Signature Event Context" and then patiently
resituates for Searle/Sarl in *Limited, Inc.*: "*The category of intention will not
disappear, it will have its place*, but from that place it will not long be able to
govern the entire scene and system of utterance" (1988, p. 59; 1982, p.
326). Derrida reiterates, "I repeat that *Sec never* adduced, from the possibility of this 'break,' the pure and simple absence of all intentionality in
the functioning of the mark that remains; rather, what it calls into question is the presence of a fulfilled and actualized intentionality, adequate
to itself and to its contents" (1988, p. 64).

Similarly, Derrida does not ignore situation/context. Derrida subjects "context" to intense and sustained scrutiny, coming to the position:

Every sign, linguistic or non-linguistic, spoken or written (in the current sense of this opposition), in a small or large unit, can . . . break
with every given context, engendering an infinity of new contexts in a
manner which is absolutely illimitable. This does not imply that the
mark is valid outside of a context, but on the contrary that there are
only contexts without any center or absolute anchoring. (1988, p. 79;
see also 1982, p.320)

For Ellis, Searle, and too many others, meanings are the realiza-
tions of the intentions of a fully conscious, present individual, a Cartesian *cogito*, the rational subject of the humanist tradition. As Searle pro-
claims, "The author said what he meant and you understand what he
said" (1977, p. 201). This dynamic is present in much of classical and
contemporary rhetorical criticism, wherein, as Dilip Gaonkar notes, a
"humanist paradigm" supports a "model of intentional persuasion" that
treats the text "as a manifestation of the rhetor's strategic consciousness" (1993, pp. 275, 277). If the meaning is not clear, Ellis suggests
turning to the speaker/author's intentions. In other words, turn to the
author as the authoritative source or origin. Derrida's texts, however,

suggest that meaning, intentions, and speakers/authors are never fully present but instead are always already inhabited by *différance* as the very condition of their possibility: "The subject, and first of all the conscious and speaking subject, depends upon the system of differences and the movement of *différance*, that the subject is not present, nor above all present to itself before *différance*, that the subject is constituted only in being divided from itself, in becoming space, in temporizing, in deferral" (1981b, p. 29; see also 1982, 1987, 1988). Far from being the fully conscious source and sovereign of discourse, the subject is the ongoing effect of social discourses, a product constituted within the matrix of linguistic and material social practices. In this sense, the subject is not a content, but a performance, a happening born, existing, and transformed in social discourses.

To conceptualize the formation of subjects and identities that do not already exist as egological selves is a difficult task.[12] What does such a subjectivity look like in practice? The practices of radical environmental groups offer a glimpse. Environmental justice groups construct selves and identities through the performance of rhetorical/social practices situated in place. To recall KFTC's tactics (Chapter 4), KFTC's identity as an environmental justice group is dependent neither on an essentialized characteristic of preconstituted group members nor on the positing of an essentialized conception of nature. Rather, the identity of KFTC emerges through the performance of tactical rhetorical/social practices designed to protect the particular place in which they are embedded from certain difficulties and threats.

This process is perhaps clearer in the transformation of Lois Gibbs from housewife to environmentalist and founder first of the Love Canal Homeowners Association and later CCHW, an "ordinary woman who, in response to crisis and challenge, transcended herself and became far more than she had been" (Levine, 1982, p. xiv). Gibbs herself describes the process:

> If I imagined a year earlier that I would be chasing Congressman LaFalce with signs, well, I wouldn't have, that's all. I am not a sign carrier. Radicals and students carry signs, but not average housewives. Housewives have to care for their children and their homes. But here I was giving press interviews, doing radio programs, and chasing a congressman, a governor, and the President with signs saying I supported him or that he was doing something wrong. Here I was literally screaming at the New York health department or the department of transportation. (1982, pp. 91–92)

The construction of Gibbs' identity through rhetorical/social practices is analogous to the construction of the identity of the Love Canal environmental justice group from "blue-collar, middle-class Americans" to activists for environmental justice (Gibbs, 1982, p. 171). As Gibbs notes, the "people of Love Canal are quite different now than they were two or three years ago. . . . [They] have changed their values, their lifestyles, and their priorities" (1982, pp. 170–171).

This construction of subjectivity has enormous implications for rhetoric, which too often has accepted the conscious intentions of strategizing individuals as sufficient explanation (in the last instance). It is not accidental, for instance, that Leff studies speeches by the likes of Lincoln and Edmund Burke. Three names, Marx, Nietzsche, and Freud, mark the disruption of a metaphysics of presence and the rewriting of the rational, conscious subject. In particular, Freud's concept of the unconscious cancels the possibility of the subject as a Cartesian *cogito*. However, if Freud thought of his work as a Copernican Revolution, many in rhetoric and communication studies still inhabit a Ptolemaic system. As Biesecker notes, the subject in rhetoric "is conceived as a consciousness, an 'I' which thinks, perceives and feels, an 'I' whose self-presence or consciousness to itself is the source of meaning" (1989b, p. 123). Rhetoric's commitment to consciousness and rationality is such that the unconscious as what one is unaware of is captured through self-reflexivity. This misses the unconscious. As Elizabeth Grosz puts it, there is a rift between the conscious and the unconscious such that

> the unconscious is not a submerged consciousness, a rational system that is somehow invisible; it is an entirely *other* form of reason, logic, and pleasure, one not reducible to those available to consciousness. It undermines the subject's conscious aspirations by its symptomatic intrusions in behaviour which are uncontrolled by, and may even be unknown to, consciousness. (1990, p. 10; see also Derrida, 1988, pp. 73–76; Spivak, 1976, pp. 33–35)

The transcendental conscious ego, the great speaker, the voice of origin, Lincoln or Burke, cannot guarantee, cannot underwrite, its meanings/ intentions.

When it comes to writing and speaking, practices of collaboration complicate notions of an undivided subject and conscious intentions. When Searle, in his "Reply to Derrida," acknowledges debts to D. Searle and H. Dreyfus, Derrida notes, "Then the 'true' copyright ought to belong to a Searle who is divided, multiplied, conjugated, shared. What a complicated signature!" (1988, p. 31). Although one may be tempted to dismiss Derrida as playing (though for Derrida play, the nonserious, and so on are exactly what we must not dismiss [1988]), questions about the subject, meaning, and intentions are writ large in the age of telecommunications. This is evident in Kathleen Jamieson's exploration of the effects of television on presidential rhetoric, which is also an un-intentional deconstruction of the traditional rhetorical subject. If we want to know the meaning of Ronald Reagan's First Inaugural Address, do we ask Reagan, who delivered it? Or Peggy Noonan, the principle speech writer? Or the many others who advised, commented, and edit-ed?

Some may argue that the rhetorical criticism of those who perform close textual analysis, by shifting focus to the text, escapes the intention-al fallacy. Although the hermeneutic character of such analysis opens possibilities for contextualizing, I think intentionality still too often gov-erns the scene of criticism. The work of Stephen Browne provides illus-trative examples.[13]

Browne's reading of Daniel Webster's eulogy of John Adams and Thomas Jefferson is framed in terms of intentionality (and transmis-sion): Webster, a Whig architect of American memory, intended to transmit cultural knowledge, specifically the message "get wisdom, speak it, and act on it" (1997, p. 42). Browne explicitly states his task in a manner that reduces the text to Webster's intended messages: "We can ask how that message gets textualized in the form of Webster's eulogy" (1997, p. 42). Browne also reads Webster's *Plymouth Rock Oration* through Webster: through his purpose, his strategic intentions, his assumptions, his meanings, his Whig politics, his vested interests, his desire to control the metanarratives of his culture (1993).

Browne's brilliant and paradigmatic close reading of Edmund Burke's *Letter to a Noble Lord* displays similar tendencies. Faced with a text that he describes as "tumultuous and seemingly out of control," that "does not appear to command an order at all" (1988, pp. 226, 221), Browne seeks recourse to biography and personal letters to ferret out the intentions that guide the subtle strategic design. A telling line from a personal letter and biographical details such as Burke's Irish birth, the Catholicism of his mother and wife, his spurning of the law and his fa-ther in favor of a literary and investment career, and his frustrating po-litical career as an opposition leader serve as prelude to Browne's study of the *Letter*'s textual dynamics. The prelude reveals Burke's bitter, suspi-cious, old "new man" character, which, in turn, reveals purpose: to de-fend the general principles of British constitutional history and his per-sonal honor from the contagion of the French Revolution. Intention, in turn, explains form (epistolary), tone (rage), and the particulars of the textual dynamics.

Browne's close readings, then, are infused with a humanist ideology that both motivates and limits the textual analyses. Essays by Gaonkar and Jasinski suggest that such a limitation is typical of this method of rhetorical criticism. Interrogating the emerging area of rhetoric of science as exemplary of contemporary rhetorical studies, Gaonkar describes a practice committed "to an agent-centered model of intentional persuasion" that "invariably reads the text as a manifestation of the author's conscious design" (1993, p. 277). Focusing on Campbell's early essays addressing Darwin's *The Origin of the Species* (1975, 1986, 1987), Gaonkar concludes, "Campbell's analysis assumes that Darwin knew exactly what he was doing and that his textual practices were intentional and premeditated" (1993, p. 280). Jasinski comes to similar conclusions about the prevalent practice of close rhetorical readings. Jasinski also discerns the search for purpose (intention) as the governing principle of rhetorical criticism. One consequence is that intention determines critical contextualization, reducing the context to a passive container or a way of uncovering authorial purpose. Jasinski's review of essays by Black, Leff and Mohrmann, and Lucas leads him to summarize the assumptions of the instrumental tradition (which includes close textual analysis) as follows: "a mode of contextualization that assumes situational stability, a sense of agency that assumes that intentions are unambiguous, fully present, and capable of directing textual production, and a sense of the text that assumes its coherence and its ability to represent authorial intention fully and without significant distortion" (1997, p. 210). Such assumptions lead not only to a certain blindness in rhetorical criticism, they also close off possibilities for the criticism of the rhetoric of a mass-mediated public sphere.

Perhaps in a period (the 18th and 19th centuries) dominated by the spoken and printed words of single speakers/authors, criticism guided by authorial intention may be the most effective method of interpreting public address. Jasinski compellingly contests such an admission by comparing Lucas' study of the Declaration of Independence with Fliegelman's textual analysis (1997, pp. 210–211; see also Hariman, 1997; Charland, 1997). In an ironic example, Campbell's (1990a, 1990b) recent attempts to read *The Origin of Species* intertextually has undermined his earlier intention-based readings of Darwin's 19th-century text (Gaonkar, 1993, pp. 280–282). Still, an intention-based study would seem to be arguably justifiable for certain historical periods. For example, even if one wants to argue that the Cartesian "I" or Lockean individual and his/her intentionality are fictions, they have been (and still are) fictions with rhetorical force and effectivity. For instance, the fictions of the individual, the autonomous author, originality, and genius are legal realities institutionalized in copyright law (Rose, 1993). Such an approach, however, is certainly not sufficient for interpreting the public discourse of our postmodern period, when contexts have been destabilized, agency and intentionality undermined, and text(s) and representation questioned. To be clear, though, as I have already mentioned in discussing Derrida on intentionality, the decentering of agency and intentionality is not the dismissal of these concepts. Agency and intentionality still must be accounted for, but in a manner that recognizes how they are forged in the complex conflux of commercial, legal, property, philosophical, and literary discourses.

The limitations of this method of rhetorical criticism are not a necessary condition of close textual analysis. The criticisms of Roland Barthes (1972, 1974, 1977) offer a method of close reading not propped up by what Leff himself terms "dubious assumptions about the role of agency and textuality in the rhetorical process" (1997, p. 131). Practicing a textual criticism that eschews the authority of the autonomous author over the discrete work, Barthes performs readings of advertisements, short stories, photographs, fashion, and so on that are not dependent on recourse to a psychological profile to produce purposes that govern the play of the text. Such analyses, instead of suturing the text into a unified work with the thread of intentionality, lay bare the irreducible polysemia that exceeds intention and escapes the unity of meaning.

The preceding discussion of intentionality and what that entails illustrates the tension between a transmission model versus dissemination/broadcasting. If Derrida's project is an unfolding of electronic communication, then an exploration of image events makes explicit the need to rethink communication from within the problematic of dissemination/broadcasting and renders commonsense notions of intention, sender (speaker, author), and audience problematic.

"War in the Woods," which is, in the age of telecommunications, a typical example of rhetoric in a mass-mediated public sphere, radically explodes the constituent elements of a transmission or instrumental model (whether described as S-M-C-R or speaker–text–audience). Earlier readings of this news report point to the ambiguity of its meaning. Yet we cannot follow Ellis' advice or Browne's methodology. We have no recourse to the speaker and his or her intentions. Instead, we are confronted by a cacophonous conglomeration of conflicting "voices." There are the environmental activists, a diverse group unified as members of Earth First!. Through their actions and words they can be understood as speaking for Earth First!. However, not only is the meaning of "body language" extraordinarily ambiguous, the report also presents the founding voices of Dave Foreman and Mike Roselle. Further, as a "disorganization," Earth First! explicitly rejects the possibility of speaking with a unified voice.

The divided, multiplied, conjugated, shared voice of Earth First! is joined and contested by those of Warren Olney of the National Forest Service, the angry logger, the hurt worker, the nurturing woman, Jay

Hair as the voice of the National Wildlife Federation and a ventriloquist for the entire mainstream environmental establishment, reporter Ken Kashiwahara, anchor Peter Jennings, the camera people, the report's writers and editors, the producers, and the owners. The news report's lack of a classical or realist narrative code prevents closure and the privileging of any particular voice. In addition, the co-presence of verbal and visual discourses divides the text, frustrating fullness for any particular discourse and undermining any voice's intentions and pretentions of presenting a unified meaning.

The complicated signature, then, renders futile any attempt to explain the meaning of a text by turning to *the* author/speaker. Rhetoric in a telecommunication age makes apparent the practical implications of Derrida's critique of intentionality and the subject. Also, Derrida's insistence on the irreducible polysemia of meaning no longer seems perverse in a context of mass-mediated electronic texts that not only can be read without the author's guarantee but must be so read, cut off, as they are, "from *consciousness* as the authority of the last analysis" (Derrida, 1982, p. 316). So while reading sentences like "Bob had a drink after work" through the lens of common sense may afford one the comfortably dim view that "all interpretations would lead to the same conclusions" (Ellis, 1991, p. 219), trying to read image events or "War in the Woods" through the distorting lens of common sense is likely to leave one uncomfortable, if not lost in the *Schwarzwald*.

The fragmentation of the author transforms the text as work into the text as a tissue of differences/signs that escapes any binding (Derrida, 1981b, pp. 33, 59–60; Barthes, 1986, pp. 56–64). What Barthes noted long ago in his essay "Death of the Author" (first published in 1968) reads like a description of the text "War in the Woods": "We know now that a text consists not of a line of words, releasing a single 'theological' meaning (the 'message' of the Author-God), but of a multi-dimensional space in which are married and contested several writings, none of which is original: the text is a fabric of quotations, resulting from a thousand sources of culture" (1986, pp.52–53).

The point of this deconstructive reading of the transmission model is not to privilege a different constituent element. Biesecker makes this very clear when she refuses to valorize either the situation (Bitzer) or the subject/speaker (Vatz), but instead offers *différance* as a nonoriginary origin, the space that makes rhetoric possible (1989b, esp. pp. 112–115). Rather, it is to highlight the idea that no constituent element can guarantee that communication (as the transmission of meaning) takes place. Instead, the taken-for-granted transmission is only a possibility within the horizon of dissemination and that today, in a postmodern age of telecommunications, instead of reading with the assumption of transmission it is more important to read within the field of dissemination. Such a perspective prevents us from appealing to the authority of a transcendental signifier, be it God, Nature, Author, Text, or Audience. It also, however, prevents the simple discarding of dominant concepts. For example, instead of doing away with the transmission model, I am trying to displace the chain of hierarchical binary oppositions that govern Western metaphysics: transmission/dissemination, presence/absence, immediacy/mediation, speech/writing, author/audience, text/audience, communication/miscommunication, reason/emotion, culture/nature, human/animal.

TEXTUAL TRIBULATIONS, CONTEXTUAL CONUNDRUMS

In other words, the text cannot replace the subject as authority and guarantor of meaning. With the poststructuralist/postmodern decentering of the subject, there has been a move in rhetoric to appeal to the authority of the text. This is most clear in the turn to hermeneutics (Hyde and Smith, 1979) and Leff's close reading or textual criticism, where "the initial locus of interest becomes the finished text rather than the person who intends to make one" (1992, pp. 223–224). I will focus on Leff's textual criticism because of its importance in the discipline. As Gaonkar notes, Leff represents one of "two dominant contemporary strategies for conceptualizing rhetoric" (1990, p. 290).[14] For the purposes of my critique in this section, I am granting the ideal form of textual criticism, which, as Leff suggests, does not seek refuge in the subject and his or her intentions. That such an ideal form is difficult to practice is not surprising given the history of the relationship between text and author since the early 1700s. As Rose explains in *Authors and Owners,* the concept of a discrete text "was underwritten by the notion of originality, which was in turn guaranteed by the concept of personality" (1993, p. 128).

For Leff, the rhetorical work/text[15] is "a field of action unified into a functional and locally stable project . . . a concrete whole—a whole that assigns meaning to a region of shared public experience and solicits an audience to embrace the meaning it constructs" (Leff and Sachs, 1990, p. 255). A text is unified, stable, concrete, whole, and dictatorial—it "assigns meaning" to public experience that an audience is then free to embrace. In short, a text is a "touchstone" (Leff and Sachs, 1990, p. 269), an object of worship that an elite cadre of critics,[16] with excruciating effort and exquisite skill, decipher for the masses in order to reveal the Truth.

I would suggest, following Barthes (1974, p. 20) and McGee (1990, p. 279),[17] that the text as touchstone has been shattered into fragments. We are left with piles of gravel that can be used for building roads among and within disparate interpretive communities. Roads marked by the traffic of everyday life. Roads that are not venerated but used, misused, worn out, rebuilt, replaced, sometimes abandoned or forgotten. Roads that always exceed intentions, that are never merely instruments for transport but also enable the formation of new communities, cultures, meanings, and identities. Roads negotiated by many different travelers: pilgrims, explorers, commuters, workers, leaders, wayfarers, thieves, tourists, shoppers, children, nomads, charlatans, peddlers, guides, and wanderers. Roads with maps that are useful but not definitive.

Texts experienced not as finished products but as arbitrary, contingent constructions—unfinished, unstable, overflowing, without integrity. For example, in this assemblage, my text could be said to be "image events," but its manifestations are diverse: the direct actions performed, preserved, and presented by environmental groups; part of the array of rhetorical/political tactics of radical grassroots environmental groups; the ABC News story "War in the Woods"; the first couple of action images the audience is likely to use to make sense of the news story.

Which is the text? All of the above and more. A text is always an intertext constituted in the space of the discursive—traversed, intersected, and constructed by historical conditions, reading practices and formations, ideologies, politics, and competing discourses (Morley, 1980b, p. 171; Chang, 1987, pp. 654–657). If Leff's project heralds "the return of the text to the center of rhetorical analysis" (Campbell, 1990, p. 249), decentering the text remains a critical task.

The privileging of the text reinforces an unfortunate tendency in rhetoric to think of the audience as "a self-evident, if not altogether banal category" (Biesecker, 1989b, p. 122). If not dismissed as a mob, the audience is posited as a collection of humanist individuals, essentially rational subjects. This is most clear in Perelman's competent and reasonable universal audience but is also implicit in the project of close reading of exemplary oratorical texts that "invite" an ideal universalized audience or, as Condit puts it, "misuniversalize" a dominant, elite, upper-class audience of close readers (1990b, pp. 330–337). The positing of such an audience strips a text of rhetorical force and reduces it to an aesthetic object. In the process, the rhetorical critic misunderstands the text as rhetoric and produces work that itself becomes art, not rhetorical interventions. For example, Leff's "virtuoso critique" misuniversalized the "invited" Northern audience and thus is unable to understand why Lincoln's rhetoric failed to reconcile North and South (Condit, 1990b, p. 336).

My close reading of ABC's news report "War in the Woods" (if not a touchstone, perhaps a representative anecdote [Leff, 1992, p. 229]) misuniversalizes the literate critic's perspective and thus fails to understand the rhetorical possibilities of image events in a heteroglossic public sphere traversed by competing discourses and peopled by disparate audience formations. Similarly, the closed world of the dominant ideology thesis rests upon misuniversalizing preferred readings.

My brief turn to audience research (qualitative, not quantitative) functions to decenter the text, but not to instantiate the audience as the privileged site of meaning. In this move I am following the better cultural studies audience research[18] and, with some misgivings, McGee's programmatic response to Leff's valorization of the text. In a much-needed antidote to Leff's textual myopia, McGee offers a contextual critical rhetoric laced with a postmodern sensibility. McGee rejects Leff's finished, stable text as an inappropriate model for rhetoric in a postmodern world. (Warnick, 1992, also makes this point.) Instead, McGee dissolves texts into contexts, "*discursive fragments of context*" (1990, p. 287). In McGee's story this fragmentation of text is related to the fragmentation of the audience (1990, pp. 279, 281, 284–285),[19] a shift from presumed homogeneity to presumed heterogeneity. This shift produces a shift in the task of text construction from authors to the audience: "*text construction is now something done more by the consumers than by the producers of discourse*" (McGee, 1990, p. 288, emphasis in original).

McGee's analysis runs into some problems here. Condit hammers him for suggesting that creative internal decoding is equivalent to text construction (1990b, p. 340). In her criticism Condit is missing the point that people are not creatively decoding finished texts (Fiske's argument), but that audiences, immersed in textual fragments, by necessity are constructing texts (meanings) out of the floating fragments of discourse traversing the social world, instead of having meaning foisted upon them by the authors of paradigmatic texts. For, Condit to the contrary, it is McGee's argument for con/text construction and his highlighting of audience activity that are to be lauded.

Condit, however, is right to call McGee on his overvalorization of the audience. McGee's analysis encounters problems due to his residual commitment to a metaphysics of presence so that the audience is a collection of psychologized individuals, unity and clarity remain unproblematic goals, and the proper context is delimited and determinable.[20]

For the fragmented text McGee substitutes the audience as the source of meaning: "The unity and structural integrity we used to put in our texts as they faithfully represented nature is now presumed to be *in us ourselves*. . . . The only way to 'say it all' in our fractured culture is to provide readers/audiences with dense, truncated fragments which cue *them* to produce a finished discourse in their minds" (1990, pp. 287, 288). McGee's description here of the rhetorical process fits easily onto the transmission template: a constructor (author) can say it all (meaning) by providing (transmitting) fragments (messages, texts) to an audience that cues (invites) them to produce (decode) a finished discourse (meaning). More troubling than the model are the ontological and epistemological commitments assumed. McGee shares with Leff a metaphysical commitment to unity, wholeness, integrity, and transparent representation.

This is also evident in McGee's defining of context. While the text is an unproblematic, unified, stable, knowable object for Leff, at times context shares those attributes for McGee:

> My way of stating the case (using the concept "fragment" to collapse "context" into "text") emphasizes an important truth about discourse: *Discourse ceases to be what it is whenever parts of it are taken "out of context."* Failing to account for "context," or reducing "context" to one or two of its parts, means quite simply that one is no longer dealing with discourse as it appears in the world. (1990, p. 283, emphasis in original)[21]

Although for McGee, "taking something out of context" (1990, p. 280) is an abuse, for Derrida it is the very structural possibility of general writing (which includes speaking), of making meaning:

> Every sign, linguistic or nonlinguistic, spoken or written (in the usual sense of this opposition), as a small or large unity, can be *cited*, put between quotation marks; thereby it can break with every given context, and engender infinitely new contexts in an absolutely nonsaturable fashion. This does not suppose that the mark is valid outside its context, but on the contrary that there are only contexts without any center of absolute anchoring. (1982, p. 320)

There is no delimited and determinable context that anchors discourse in the real world. Further, even on a practical level, context is not delimitable. Rather, the heterogeneity of audiences points to the undecidability of context, what Derrida terms "structural nonsaturation" (1982, p. 310). Which context makes "War in the Woods" into "discourse as it appears in the world": a bar in Idaho, a cafe in Boston, a suburban home in Califirnia, a dorm room at the University of Iowa, an EarthFirst! meeting, or a corporate boardroom? To deny the possibility of an absolutely determinable context is not to say a text has no context. Rather, a text has partial and contingent contexts. In the examples just mentioned, the audiences share aspects of contexts: they are in an industrialized United States, they live in a world with ozone holes, they are familiar with television, and so on. Those contexts shape the contingent meanings of the text.

In a move endemic to rhetorical studies, then, McGee extends the attributes of the humanistic subject, "a consciousness, an 'I' which thinks, perceives and feels, an 'I' whose self-presence or consciousness to itself is the source of meaning" (Biesecker, 1989b, p. 123) to the audience as a collection of individuals or, considering McGee's reversal of speakers/writers and audiences/readers (1990, pp. 274, 288), a collection of authors. McGee's conventional extension is important in light of his radical reversal, for while significantly problematizing traditional notions of author and text (note Condit's disapproval, 1990b), McGee simultaneously recuperates Western culture's investment in a metaphysics of presence through his conceptualization of the audience as the site where unity, integrity, faithful representation, in short, the transmission of meaning, are secured. McGee, after overturning the author/audience hierarchy, needed, instead of replacing it with an audience/author hierarchy, to displace the hierarchy through a deconstructive reading of the audience.

Such a deconstructive reading necessarily starts with Derrida's deconstruction of the subject, touched upon earlier, as a being inhabited by *différance,* which traverses, marks, and/or is the condition of possibility for the sign, signified, signifier, presence, subject, consciousness, system of the same, Being, and beings. *Différance* "is the non-full, non-simple, structural and differentiating origin of differences ... the movement according to which language, or any code, any system of referral in general, is constituted 'historically' as a weave of differences" (Derrida, 1982, pp. 11, 12). For the self-conscious subject inscribed in language, "*différance* would be not only the play of differences within language but also the relation of speech to language, the detour through which I must pass in order to speak, the silent promise I must make" (Derrida, 1982, p. 15).

If subjectivity is not linked to an immutable essence, but rather is an effect of *différance,* rhetoricians can no longer "presume the presence of an audience that finds, in any rhetorical situation, its ontological and epistemological foundation in the notion of a sovereign, rational subject" (Biesecker, 1989b, p. 123). Without such a presumption, the audience cannot be the source and site of unity, integrity, and meaning in a finished discourse.

In sum, a turn to the audience is a justifiable response to the dual tyrannies of the author and text, but the audience must not remain an unexamined category that reinstantiates tyranny. This is both a theoretical and a practical danger. McGee's elevation of audience/readers is one theoretical example, as is Barthes' earlier elevation of the reader in his groundbreaking "The Death of the Author":

> A text consists of multiple writings, proceeding from several cultures and entering into dialogue, into parody, into contestation; but there is a site where this multiplicity is collected, and this site is not the author, as has hitherto been claimed, but the reader: the reader is the very space in which are inscribed, without any of them being lost, all the citations out of which a writing is made; the unity of a text is not in its origin but in its destination. (1986, p. 54)

In practice, both quantitative and qualitative audience research too often mistake the necessary detour through the audience as the way to Truth. As in the examples of Barthes and McGee, the audience becomes the site/source of meaning, of answers.[22] Derrida's deconstruction of subjectivity checks the idealization of the audience, which then becomes recognized as an effect of *différance,* language, competing social and economic discourses, texts, and histories. Such a perspective rejects any essential or fixed identity of the audience, whether such an identity be attributable to ontological or demographic characteristics, and instead considers the audience to be "bundles of practices," which recognizes the multiplicity of decodings available, the ways histories and contexts traverse and constitute audiences and readings, and the fluidity and contingent character of both audiences and readings so that the "same" audience may offer different readings of the "same" texts (Chang, 1987, pp. 659–662).

Yes, meaning happens at the site of the audience, but it is marked not by unity, integrity, faithfulness, and finality, but by conflict, contradiction, complexity, and contingency, the result of negotiations between audiences, texts, authors, and contexts wherein none of these constituent elements is self-identical or originary. For example, the audience, embedded at a multiplicity of sites, is not cued in by the fragment or compilation of fragments titled "War in the Woods" "to produce a *finished* discourse *in their minds*" (McGee, 1990, p. 288; emphasis added). Rather, audiences without "unity and structural integrity" engage in a social–historical process that produces many meanings, in principle an irreducible polysemia, a dissemination. The image events of "War in the Woods" mean differently in a bar in Wyoming, a cafe in Seattle, a logging community in Idaho, a Sierra Club meeting in Washington, DC, a home in Iowa, and so on. In dissemination, there is no site that collects the irreducible multiplicity of meanings.

CHAPTER 7

RHETORIC AND SOCIAL CHANGE IN A POSTMODERN CONTEXT

In these meditations, these fragments, I have been attempting to perform a postmodern critical rhetoric. McKerrow, in his programmatic essay "Critical Rhetoric" (1989), codifies earlier attempts by rhetorical theorists (McGee, Wander) to conceptualize rhetoric as a practice for critical intervention in the exercise of social power. McKerrow's essay, however, represents codification with a twist. McKerrow's antecedents relied on versions of Marxism and the Frankfurt School's Critical Theory, but McKerrow significantly shifts grounds to poststructuralism, especially the work of Michel Foucault, Ernesto Laclau, and Chantal Mouffe, in order to explore the possibilities of rhetoric as critical intervention in a postmodern[1] world.

This momentous shift opens up opportunities for intervention by critical rhetoricians. Unfortunately, besides those who simply dismiss critical rhetoric, most responses are limited to one of two tacks. The first is to criticize McKerrow for being, like most of us, suspended between modernity and postmodernity (Hariman, 1991; Biesecker, 1992). Although the criticisms are often insightful, they are neither surprising nor particularly productive, especially when one considers that the very term "postmodern" would seem to preclude theoretical purity. The second approach seeks to supplement critical rhetoric with terms and theorists from classical rhetoric. Charland (1991) offers us *phronesis* and Aristotle, Ono and Sloop (1992) proffer *telos* (admittedly amended), and Clark (1996) submits service (*ophelia* and *douleia*) and Isocrates. Although all of these theorists are adding to the conceptualization of critical rhetoric, they are leaving largely unexplored the possibilities presented by McKerrow's turn to poststructuralism. I think a poststructuralist/ postmodern perspective offers distinct advantages. In order to explore those advantages I respond to criticisms of critical rhetoric and expound upon the benefits for critical rhetoric of the turn to discourse and poststructuralist subjectivity within a postmodern problematic. From this perspective, the primary task of the critical rhetorician is context construction, a task with both synchronic and diachronic dimensions. Performing this task, I close with an attempt to read the discourse of industrialism otherwise through a historical revalorization of the contemporary epithet "Luddite."

DISCOURSE, REALITY, AND POLITICS

Critical rhetoricians and poststructuralists face two major attacks. The first is that they "collapse the distinction between discourse and the real" (Cloud, 1994, p. 154; see also Eagleton, 1991). For instance, in a largely laudatory appraisal, Stuart Hall voices the concern about *Hegemony and Socialist Strategy* that the "book thinks that the world, social practice, is language" (Hall, 1986, p. 57). The accusation that poststructuralists retreat into language and leave the "real" world behind is based on erroneously equating discourse with language, for poststructuralism suggests that discourse is material and includes within it the linguistic and the nonlinguistic.[2] Laclau and Mouffe use the term "discourse" to emphasize "that every social configuration is *meaningful*. . . . In our interchange with the world, objects are never given to us as mere existential entities; they are always given to us within discursive articulations. . . . Outside of any discursive context objects *do not have* being; they have only *existence*" (1987, pp. 82, 85).

Of course a tree exists, but a tree is not just a tree. It is firewood, a god, shelter, a source of food, or artistic inspiration depending on the discursive context. To use an example from environmental politics, yes, toxic waste dumps exist. Their existence is not in question, but what they *mean* is the site for political struggle. Within the hegemonic discourse of industrialism, toxic waste dumps are "the price of progress," the normalized cost of economic growth, and the people affected by their siting need to sacrifice for the common good. Environmental justice groups are struggling to articulate an alternative discourse of environmental justice that contests this meaning of toxic waste dumps and rearticulates them as examples of class discrimination, institutional racism, and corporate colonialism that expose the limits of and challenge the discourse of industrialism, thus expanding the spaces for political struggle and resistance.

The second major attack critical rhetoricians and poststructuralists face is that the discursive turn, because it eviscerates the critic's privileged position, neuters critique. This leads to what critic Cloud describes as "the evacuation of the critical project" in favor of "the aestheticization and depoliticization of political struggle" (1994, pp. 159, 157). This charge points to a pervasive fear that the abandonment of a logic of a priori necessity, essential identities, and foundations (laws of History, economic determinism, universal class, etc.) will make politics and social critique impossible. If one shares, as I do, the conviction that only participation in a critical project justifies our privileged position in academia, then this is not an inconsequential fear. Further, as discussed in Chapter 4, such a fear is endemic to our postmodern moment (see Habermas, 1987; Harvey, 1989) and is perhaps most revealingly expressed by Harvey when he complains that "postmodernism, with its emphasis upon the ephemerality of *jouissance,* its insistence upon the impenetrability of the other, its concentration on the text rather than the work, its penchant for deconstruction bordering on nihilism, its preference for aesthetics over ethics, takes matters too far. It takes them beyond the point where any coherent politics are left" (1989, p. 116). I read this fear as a clue to the crucial question of the possibilities of politics and rhetoric in a postmodern age.

The poststructuralist response to this fear of a retreat from the political is that the abandonment of ultimate foundations and the widening of the field of undecidability expand the field of politics (Laclau, 1993b, p. 280; see also Butler, 1992), that the subversion of structural laws by contingency creates the very possibility of radical politics (Laclau, 1990, p. 46; see also Laclau and Mouffe, 1985). If society and history are understood as the necessary unfolding of Reason getting to know itself (Hegel) or the necessary development of the laws of History (Marx), politics is reduced to discovering the action of a reality external to itself and humans are reduced to spectators or actors in a scripted play written by Reason or History (or some other essential ground). If, however, society is understood as groundless, politics becomes ontological as the name of that process through which social agents in part construct their own world (Laclau, 1993a, p. 341; 1993b, p. 295). As Laclau explains,

> Abandonment of the myth of foundations does not lead to nihilism, just as uncertainty as to how an enemy will attack does not lead to passivity. It leads, rather, to a proliferation of discursive interventions and arguments that are necessary, because there is no extradiscursive reality that discourse might simply reflect. Inasmuch as argument and discourse constitute the social, their open-ended character becomes

the source of a greater activism and a more radical libertarianism. Humankind, having always bowed to external forces—God, Nature, the necessary laws of History—can now, at the threshold of post-modernity, consider itself for the first time the creator and construc-tor of its own history. (1993a, p. 341)

In short, the discursive turn expands the possibilities and importance of politics *and* rhetoric. Within a discursive frame rhetoric is no longer a *techne* or instrument in the service of Truth (be it Platonic or Marxist), but rather, becomes constitutive of any social or political collectivity.

To take examples from environmental politics, if the foundation of society does not rest on God's granting dominion over nature to "man," the clear-cutting of forests becomes a political issue. If progress is not accepted as the grand narrative of industrial society, then toxic waste dumps become the sites of rhetorical and political struggle.

The postmodern critique of essentialism and foundations not only expands the field of politics in regard to theory but recognizes both that the essentialized identity of the working class has failed in the wake of the dislocations of disorganized or late capitalism and that the "new so-cial movements" (Melucci, 1980, 1985; Offe, 1985; Touraine, 1985) have emerged based on the dispersion of subject positions and the pro-liferation of struggles in the contemporary social field. The various struggles of environmental, feminist, anti-racist, and anti-nuclear peace groups share the central characteristic "that an ensemble of subject po-sitions linked through inscription in social relations, hitherto considered as apolitical, have become loci of conflict and antagonism and have led to political mobilization" (Mouffe, 1992, p. 372).

CRITICAL RHETORIC, SUBJECTIVITY, AND THE PLACE OF THE CRITIC

Although some decry critical rhetoric's "easy adoption of poststruc-turalist ideas" (Cloud, 1994, p. 159), actually the adoption has been hes-itant and incomplete. McKerrow's vestiges of modernism lead to incon-sistent deployments of postmodern theories that hamper critical rhetoric. In this section I explore how poststructural theorizing of the subject offers us a chance to more fully formulate the task and agency of the critical rhetor.

The problems in McKerrow's formulation of critical rhetoric result from his retaining certain modernist assumptions while attempting to move toward an orientation appropriate for the "problematic of post-modernism" (McKerrow, 1991a, p. 76). For example, McKerrow proposes that "a critical rhetoric seeks to unmask or demystify the discourse of power" (1989, p. 91; see also pp. 92, 98, 100; and McKerrow, 1991b, p. 250). As Biesecker astutely notes, Foucault and other poststructuralists would be dubious of the "claim that it is out of shared knowledge delivered over to the audience by the rhetorician that the collective desire and power to contest forces of domination arises" (1992, pp. 352-353). Further, the project of demystification posits the critical rhetor as a modern, rational subject observing the social scene from a privileged stance and enlightening the masses mired in false consciousness. Indeed, commentators have taken McKerrow to task for creating the critic as "a supposed autonomous agent . . . a disembodied thinker having no identifiable social location" (Ono and Sloop, 1992, p. 51; Hariman, 1991, p. 68).

In short, far from displacing the modern critic, McKerrow seems to have resurrected just such a critic. I want to suggest, however, that sympathetic critics like Biesecker, Ono and Sloop, and Hariman have not been charitable in reading the strand of modernism that runs throughout McKerrow's formulation of critical rhetoric as the whole tapestry. A more productive path would recognize the modern–postmodern tension in the essay and emphasize the postmodern strands in McKerrow's heuristic image of critical rhetoric.

Notwithstanding his call to demystify and "expose the discourse of power in order to thwart its effects in a social relation" (1989, p. 98), in other sections McKerrow, following Foucault, eschews analyses that "have as their motive the 'demystification of ideologically distorted belief systems'" (Fraser, quoted in McKerrow, 1989, p. 97) and argues that in critical rhetoric the "orientation is shifted from an expression of 'truth' as the opposite of 'false consciousness' (and away from the naive notion that laying bare the latter would inevitably move people toward revolution on the basis of a revealed truth)" (1989, p. 99; see also pp. 93, 100, 104).

Although McKerrow's conceptualization of critical rhetoric tends to call forth a modern subject and traditional notions of rhetor, audience, and critic, despite his own critique of the traditional model of rhetoric (1989, pp. 100–101), McKerrow does painfully pursue poststructuralist notions of subjectivity. He denies the possibility of preconstituted subjects and instead speaks of fractured subjects who are articulated in the constitutive process of hegemony (1989, pp. 105, 94), with hegemony understood as a discursive practice encompassing both language and social practices.

Recognizing the incompleteness of his theorizing of the subject in critical rhetoric and eager to respond to charges that he instantiates a modern subject (Hariman, 1991; Ono and Sloop, 1992), in a later essay

McKerrow explores the possibility of the subject in critical rhetoric (1993). Relying on Foucault and, to a lesser extent, Derrida, McKerrow proffers a historically grounded subject constituted within and transformative of a matrix of social practices (1993, pp. 61, 60), a subject "decentered and viewed as a form rather than a substance, as the intersection of truth rather than the being that finds truth" (1993, p. 64). This discursive perspective highlights the critique of the essentialist subject as "an originative and founding totality" and instead offers "'subject positions' within a discursive structure" (Laclau and Mouffe 1985, p. 115). It bears repeating, however, that a subject is not simply interpellated by one discourse; rather, a subject is constituted as the nodal point of a conglomeration of conflicting discourses, a position that leaves room for agency but not the free will of a preconstituted subject. As Mouffe expounds,

> We can thus conceive the social agent as constituted by an ensemble of "subject positions" that can never be totally fixed in a closed system of differences, constructed by a diversity of discourses among which there is no necessary relation, but rather a constant movement of overdetermination and displacement. The "identity" of such a multiple and contradictory subject is therefore always contingent and precarious, temporarily fixed at the intersection of those subject positions. (1993, p. 77)

Given this rough-hewn portrait of postmodern subjectivity, which I tried to flesh out in the last chapter by exploring how environmental justice groups construct selves and identities through the performance of rhetorical/social practices situated in place, what is the task of the critic? Despite certain longing intimations to the contrary, it certainly is not to reveal the truth and enlighten the masses. Rather, the critic, no longer a disembodied, universal thinker but instead an embodied, specific intellectual (McKerrow, 1989, p. 108; 1993, p. 62; Foucault, 1980, pp. 126–133) "constituted in and through the same contingent social practices of those for whom the critique is performed" (McKerrow, 1993, p. 62), is to invent a text and become "arguer or advocate for an interpretation" (McKerrow, 1989, pp. 108, 101; 1993, p. 62) within the larger war of interpretations that constitutes the public sphere. Although the choice of the word "invent" invokes potent rhetorical roots, it also carries unfortunate modernist connotations of discovery and the lone genius. Some of these connotations are evident when McKerrow describes "the role of the critic as 'inventor'—interpreting for the consumer the meaning of fragments collected as *text* or *address*" (1989, p. 101).

"Assemble" is a better choice than "invent," both because it can trace its roots to McGee's "fragments" essay (1990) and because it brings up echoes of *bricolage,* thus positioning the critical rhetorician less as an inventor and more as a *bricoleur* (Spivak, 1976, pp. xviii–xx). Whatever the choice of word, all point to the task of the critical rhetorician being akin to the task of Foucault's specific intellectual:

> The essential problem for the intellectual is not to criticise the ideological contents supposedly linked to science, or to ensure that his own scientific practice is accompanied by a correct ideology, but that of ascertaining the possibility of constituting a new politics of truth. The problem is not changing people's consciousnesses—or what's in their heads—but the political, economic, institutional regime of the production of truth. (1980, p. 133)

In short, the task is one of context construction, changing not people's ideas but the conditions of possibility for thinking, of transforming terministic screens, intellectual grids, paradigms. This is a particularly crucial task in a postmodern age.

SETTING THE CONTEXT FOR SOCIAL MOVEMENTS

Radical environmental groups are operating in what can be described as a postmodern social field. The characteristics of such a field offer significant advantages to radical environmental groups, not the least of which is the distrust of grand narratives like progress and the valorization of the local. There are, however, political disadvantages to a postmodern social field. Key among these is time–space compression (Harvey, 1989), which results in what Katherine Hayles (1990) terms the "denaturing of context."

If modernism in its embrace of progress sought a break with history and tradition, at least the very need to seek a break with the past implied a connection. Now, historicity has been so weakened, as Gallup Polls, our interactions with students, and Disney's venture to pave over our memories of the Civil War with a theme park suggest, that to "live postmodernism is to live as schizophrenics are said to do, in a world of disconnected present moments that jostle one another but never form a continuous (much less logical) progression" (Hayles, 1990, p. 282).

The modern rationalization and homogenization of space under the systematized organization of capital, complemented by the postmodern compression of space that leads us to metaphorically conceive of our world as a "global village" or "spaceship earth," work to obliterate particular places. This puts radical environmental groups at a distinct disadvantage, for they are "generally better at organizing in and dominating *place* than they are at commanding *space*" (Harvey, 1989, p. 236).

With the combination of "the advance of remorseless time and space, the past becomes lost and falls into nothingness" (Berger, 1984, p. 37). In other words, the postmodern time–space compression denatures context, so "that contemporary Americans live 'within the context of no context'" (Hayles, 1990, p. 272). One consequence is that radical environmental groups and their actions are understood not in the context of a rich history of social activism on behalf of the victims of "improvements" and against the adverse effects of progress in the United States and elsewhere, but as decontextualized, isolated, commodified images indistinguishable within a commercial system of juxtaposed images that lack internal connections. Angus and Jhally offer an incisive explanation of how this system works:

> The commercial system requires a continuous influx of new cultural commodities. Assimilation of cultural productions into this system strips them of their original context and presses them into an image-form. Without a context for interpretation of images, they all blend into an undifferentiated continuous flow, in which each individual image or set of images, has no particular significance. Thus, they succumb, whatever their intent or content, to the mainstream assumptions of the society at large which dominate the conditions of reception. (1989b, p. 13–14)

Often the only attempt at providing context is through the construction of media frames. As we saw in "War in the Woods," radical environmental groups are constructed and condemned by the media through such frames.

The political significance of the postmodern compression of time and space and the consequent evisceration of context is evident when activists such as Anne Braden of the Southern Organizing Committee for Economic and Social Justice argue that "we have to learn from history . . . [and] [w]e've got to build a regional struggle around specific local struggles" (quoted in Madison, 1993 p. 32). For oppositional groups, such as radical environmental groups, that "are relatively empowered to organize in place but disempowered when it comes to organizing over space" (Harvey, 1989, p. 303), the needs to learn from history and to make connections among local struggles point to the necessity of context control (Hayles, 1990, p. 274) in what can at best be described as a postmodern context of no context. Context construction is one of the vital tasks for critical rhetoricians. Such a task, however, is political, not historical or revelatory.

Given the disparate, contingent, and fragmentary rhetoric of radical environmental groups trying to empower themselves, to demand a redress of their grievances, and to create social movement, my task and political responsibility as a critical rhetorician is to assemble a con/text (McGee, 1990). There are two ways that critical rhetoricians can help contingently to construct and control context. The first is "to trace new lines of making sense by taking hold of the sign whose reference has been destabilized by and through those practices of resistance, lines that cut diagonally across and, thus disrupt, the social weave" (Biesecker, 1992, p. 361). In other words, the critical rhetorician can help reconfigure the grid of intelligibility so that the tactics, acts, and image events of radical environmental groups, including blocking a bulldozer, plugging a toxic discharge pipe, or smashing a machine, are not conceived as illegal acts of obstructionism, vandalism, or terrorism. Rather they can be recognized as legitimate political acts that call into question the morality and legality of acts by corporations that displace people and ravage the environment. To that end, my attempt at postmodern critical rhetoric has focused on the "illegitimate" rhetorical activity of local, subaltern groups whose tactics violate the propriety of sites and institutions that dictate who can speak, what can be said, and how it can be said (King, 1992, pp. 1–12).

Although the first way operates on the synchronic level, the second way of constructing and controlling the context, the conditions of reception and interpretation of radical environmental groups and their image events, consists of unearthing the diachronic structures of key ideographs like progress, technology, and nature. I am not suggesting conducting history as an exhaustive archaeological dig but as daring raids, search and rescue missions. As Walter Benjamin explains, "To articulate the past historically does not mean to recognize it 'the way it really was' (Ranke). It means to seize hold of a memory as it flashes up at a moment of danger. . . . Only that historian will have the gift of fanning the spark of hope in the past who is firmly convinced that *even the dead* will not be safe from the enemy if he wins. And this enemy has not ceased to be victorious" (1968a, p. 255).

Such historical analysis embeds radical environmental groups and their tactics within an important tradition of opposition to the evils and excesses of industrialism, that have been justified as the price of progress. To close this assemblage I will begin such an analysis, for the history of struggle for environmental justice provides tactical, psychological, and moral resources, in a word, hope, for the struggles of today. "Hope implies a deep-seated trust in life that appears absurd to those who lack it. It rests on confidence not so much in the future as in the past. . . . Hope does not demand a belief in progress. It demands a be-lief in justice" (Lasch, 1991, p. 81, 80).

REDRESSING PROGRESS

A Klee painting named *Angelus Novus* shows an angel looking as though he is about to move away from something he is fixedly contemplating. His eyes are staring, his mouth is open, his wings are spread. This is how one pictures the angel of history. His face is turned toward the past. Where we perceive a chain of events, he sees one single catastrophe which keeps piling wreckage upon wreckage and hurls it in front of his feet. The angel would like to stay, awaken the dead, and make whole what has been smashed. But a storm is blowing from Paradise; it has got caught in his wings with such violence that the angel can no longer close them. This storm irresistably propels him into the future to which his back is turned, while the pile of debris before him grows skyward. This storm is what we call progress.
                    —Walter Benjamin, *Theses on the Philosophy of History*

To all Croppers, Weavers &c. & Public at large.
Generous Countrymen,
You are requested to come forward with Arms and Help the Redressers to redress their Wrongs and shake off the hateful Yoke of a Silly Old Man and his Son more silly and their Rogueish Ministers, all Nobles and Tyrants must be brought down. . . .
Apply to General Ludd Commander of the Army of Redressers.
                    —Luddite leaflet

The lack of a historical context for the various struggles of environ-mental groups not only hampers public understanding of such groups but hinders the groups' understanding of each other, so that radical preservation groups and environmental justice groups regard each other with wariness, if not hostility. These groups need to recognize that their common struggle against industrialism mystified as progress.

Since the establishment of the articulation of industrialism and the elevation of progress as the key ideograph, there have been a number of attempts to read progress otherwise, to read it against the grain. These attempts provide a context for making sense of environmentalisms to-day. For example, Robert Marshall, the founder of The Wilderness So-ciety, advocated a "democratic wilderness" and "sought to link social justice and wilderness protection" by "linking protectionist objectives within a social policy framework" (Gottlieb, 1993, pp. 18, 19, 17). Through his practices as a radical forester and even in his will, wherein he divided his $1.5 million estate into three trusts for social advocacy, the promotion of civil liberties, and wilderness preservation, Marshall showed that a comprehensive critique of industrialism and progress requires a critique of both the exploitation of nature and the exploitation of people. Further, through his work Marshall made manifest his belief that the liberation of society was a necessary condition for the liberation of nature and that the liberation of nature was an essential condition for social liberation (Gottlieb, 1993, p. 15).

Instead of continuing this analysis with a man whose legacy has been silenced by his own organization, perhaps it would be more significant to revisit a group whose very name has become a pejorative word used to dismiss anyone questioning technological progress: the Luddites. When essayist, poet, and farmer Wendell Berry makes an ecological and social argument against using computers he is ridiculed as a Luddite. Business columnist Joseph Nocera uses Luddite to tar the traitor Jimmy Goldsmith, a former corporate raider turned critic of industrial capitalism: "He is not a racist like Ford or a crank like Perot. What he is, to put it plainly, is a Luddite. Anything that smaks of progress immediately makes him suspicious" (1995, p. 76). Earth First!ers who value wilderness over machinery and disable the latter in defense of the former are railed against as Luddites. Environmental justice groups seeking to plug the toilet of the industrial system, if not branded Nimbys, are labeled Luddites obstructing the progress of modern civilization. Basically, anyone suspicious of ATMs, nostalgic for the sound of human voices on the phone, wary of computers, frightened by cars, disdainful of the information superhighway, or against automation is kept in line or at least on the defensive with the label Luddite. As technology critic Robert Rossney notes, "'Luddites' itself has been enjoying a renaissance right now, because in a world full of rapid change there's a lot of demand for a label to stick on those people who don't particularly want it to happen. . . . It's invariably pejorative. You don't call someone a Luddite to compliment his wisdom in doing something the old-fashioned way" (1994, p. 3).

In popular usage, then, a Luddite is a reactionary technophobe standing in the way of progress. "Luddite" functions as a club to beat opponents with, as technothriller novelist Tom Clancy did when attacking opponents of President Reagan's Strategic Defense Initiative (Watson, 1994, p. 154). In a society where technocratic thinking—the belief that technology is a force autonomous from politics and ideology that produces inevitable social effects that societies must adapt to (Leiss, 1989; Leonard, 1993; Noble, 1977)—is endemic, Luddite is a rhetorical weapon, a devil term, used to silence those who question progress. However, when E. P. Thompson salvaged the Luddites from the dustbin of history, he was salvaging not an anarchic mob of technophobic reactionaries but rather "*a quasi-insurrectionary movement*" (1964, p. 553, emphasis in original) that through its actions presented the first collective critique of industrialism and progress. From this perspective, then, the Luddites mark the beginning of a history of groups that, through their various actions, critique industrialism and seek to read progress otherwise.

That the Luddites are worthy forebears for both environmental justice and radical environmental groups is evident in the comprehensiveness of their critique and their attempts to move the meaning of the ideograph progress. The Luddites petitioned Parliament—"between 1803 and 1806 workers raised the astonishing amount of 10,000 pounds to lobby for enforcement of antimachine laws" (Watson, 1994, p. 143)—and employed selective direct actions against the machinery of capitalists who violated the tradition of a just price and fair wage (Thompson, 1964, pp. 521–602). Through these means they demanded "a legal minimum wage; the control of the 'sweating' of women or juveniles; arbitration; the engagement by the masters to find work for skilled men made redundant by machinery; the prohibition of shoddy work; the right to open trade union combination," and the restoration of constitutional rights (Thompson, 1964, pp. 552, 547). The demands represented an attack on a notion of progress defined as: Capitalists, through free trade and the use of technology, have the right to exploit labor and the environment in pursuit of profits, regardless of the consequences for others. This version of progress was supported by a Parliament that repealed all laws regulating the wool industry in 1809 (Watson, 1994, p. 143) and made machine-breaking and secret oath-taking capital offenses (Thompson, 1964). As E. J. Hobsbawm observes, after 1660 the state's "traditional hostility to devices which take the bread out of the mouths of honest men, gave way to the encouragement of profit-making enterprise, at whatever social cost. . . . [By the 1800s] the voice of the manufacturer increasingly became the voice of government" (1952, pp. 65, 66). The Luddites recognized the dangerous implications of this new industrial progress articulated by capitalists and Parliament:

> What was at issue was the "freedom" of the capitalist to destroy the customs of the trade, whether by new machinery, by the factory-system, or by unrestricted competition, beating-down wages, undercutting his rivals, and undermining standards of craftsmanship. We are so accustomed to the notion that it was both inevitable and "progressive" that trade should have been freed in the early 19th century from "restrictive practices," that it requires an effort of imagination to understand that the "free" factory-owner or large hosier or cotton-manufacturer, who built his fortune by these means, was regarded not only with jealousy but as a man engaging in *immoral* and *illegal* practices. The tradition of the just price and fair wage lived longer among "the lower orders" than is sometimes supposed. They saw *laissez faire,*

not as freedom, but as "foul Imposition." They could see no "natural law" by which one man, or a few men, could engage in practices which brought manifest injury to their fellows. (Thompson, 1964, p. 549, emphasis in original)

As the Luddites realized, what was at stake in this meaning of progress was not only their jobs, but their communities and way of life. "They weren't attacking technology because they feared for their jobs. They were attacking concentrations of capital because they feared for their culture" (Rossney, 1994, p. 8).

These attacks, then, were a form of rhetoric arguing for an interpretation of progress that cares less about profit and technology and more about people and places. Thus, far from being "pointless, frenzied, industrial jacquerie" (Plumb, quoted by Hobsbawm, 1952, p. 57) or "the overflow of excitement and high spirits" (Hobsbawm, 1952, p. 57), Luddism names the rhetorical activity of groups of human beings attempting to produce social movement, a change in human consciousness, with regards to the ideograph progress and the hegemonic discourse of industrialism. The Luddites read industrial progress as a catastrophic storm threatening to wreck their jobs, communities, and culture and leave a pile of debris. The demands of the Luddites foreshadowed an image "of a democratic community, in which industrial growth should be regulated according to ethical priorities and the pursuit of profit be subordinated to human needs" (Thompson, 1964, p. 552).

For environmental and other social activists experiencing the birth throes of the Information Revolution while still reeling from the social and environmental effects of the Industrial Revolution, after being legally marginalized by NAFTA and GATT while being bloodied by the Wise Use movement, in short, in the wake of a renewed cultural offensive of scientism and progressivism in which the drive for total automation is promoted in the name of patriotism, competitiveness, productivity, and progress while its twin aims remain control and domination (Noble, 1984, p. 328), the Luddites' demands and goals still resonate today and offer a spark of hope for an alternative articulation of progress. It is a hope embedded in place.

The Luddites were implaced in their defense of the places of Nottinghamshire, Yorkshire, Lancashire, and Cheshire. Theirs was a tactical defense of their culture and way of life against the onslaught of industrialism in the form of "Machinery hurtful to Commonality" (Luddite leaflet, reprinted in Sale, 1995, p. 118). The invasion was enforced by the strategic power of imperial England (London), most starkly in the form of a "foreign" army of 14,400 soldiers (a force larger than the one that accompanied the Duke of Wellington to Portugal for a war against Napoleon 4 years earlier [Sale, 1995, p. 148]). The Luddites, then, were combating the still prevalent global strategy of domination that the philosopher Edward Casey, in his book *The Fate of Place*, terms "deplacialization": "the systematic destruction of regional landscapes that served as the concrete settings for local culture" (1997, p. xii), with landscape being understood not merely in terms of physical topography, but also including a people and their practices—not merely mathematical space but lived place.

Within this historical context of Luddism, the term "Luddite" can be used not only as an epithet by the proponents of industrial progress, but as a watchword for those who would resist the practices and products of industrialism. Place is the keystone to resistance to industrialism then and now. Although many tactics can be coopted, a commitment to particular places is antithetical to the discourse of industrialism on many fronts. Place interrupts the abstraction and universalization of value as exchange value, that is, money; the privileging of instrumental reason (scientific rationality) as *the* way of knowing; the conceit of a universal "man"; quantification of nature as a storehouse of resources; and the positing of infinite, absolute Newtonian space and linear time. Indeed, the history of industrialism can be read as the tale of the obliteration of local places in favor of national and international spaces. The particularities of places have been paved under the coordinates of geometric space and the ticking of chronometric time. Following the example of the Luddites, Earth First! and other radical environmental groups and KFTC, ACNag, and other environmental justice groups are resisting the industrial process of deplaciation whereby the "cosmos itself, formerly a matrix of places, has yielded to the spatial (and temporal) imperialism of the *universum* (literally, the whole 'turned into one')" (Casey, 1997, p. 199).

Earth First! consciously invokes the legacy of Luddism, publishing books under the imprimatur of Ned Ludd Books, offering a radical critique of industrialism, and adapting the tactic of destroying "Machinery hurtful to Commonality" (though ecotage is consciously nonviolent, the Luddites' actions were not). They are also linked to Luddism in their emphasis on saving places. In their image events Earth First!ers inhabit wilderness areas, turning spaces quantified as board feet of lumber into particular places. The inhabiting of trees and regions by Earth First! activists is important. As brief glimpses of camp sites suggest, Earth First!ers often live in the places they are trying to protect. They dwell in the woods. Trees-sitters live with the trees. Merleau-Ponty suggests that bodies and places share a coconstitutive relationship. Space becomes lived places through the inhabitation and movements of bodies and,

In defense of places, Earth First! activists often dwell in trees for months at a time. By living in the woods, the activists create an implaced ecocentric community.

conversely, "I already live in the landscape" (1962, p. 251; see also Casey, 1997, pp. 228–238). From this perspective, place is less a geographical entity and more an event, an experience (Snyder, 1990, pp. 25–47; Casey, 1997, pp. 334–336). By implacing their bodies in a region through burying themselves in the ground, perching in trees, hugging trees, and living in these areas until forcibly removed, Earth First!ers make a particular place. Indeed, by dwelling in nearness with nonhuman Others, they constitute an ecocentric community. Through the care of the neighbor, a tree becomes *this* tree, a forest *this* forest, a mountain *this* mountain. For example, Julia "Butterfly" Hill has lived at 180 feet in one redwood for over a year now. Through dwelling in the tree, she feels, "I have become one with this tree and with nature in a way I would never have thought possible" (quoted in Hornblower, 1998). Through inhabitation, a space that is interchangeable with many others becomes *this* place that is irreplaceable in its particularity.

This connection to place is highlighted in the names of many of the local manifestations of Earth First!. Among others, there is Northcoast EF! in California, Gunnison Basin EF! in Colorado, Tallgrass Prairie EF! in Iowa, Yellowstone EF! in Montana, Hock-Hocking Watershed EF! in Ohio, Kalmiopsis EF! in Oregon, and Allegheny EF! in Pennsylvania.

Though the connection of environmental justice groups to the Luddites tends not to be verbalized, in many ways their concerns and defenses of places coincide. Both critique the practices of industrial capital that harm workers, communities, and traditions of local cultures. Both defend places that are not wildernesses, but human habitations that have a sedimented sense of place accrued through performances of place over time. In saying no to destructive corporate practices, environmental justice groups are both challenging an industrial worldview that operates on a spatiotemporal grid that cannot even see particular places and reaffirming their commitment to *this* place made their own through the everyday practices of living. By performing rhetorical tactics situated in place, by creating a rhetoric of bodies in place, environmental justice groups have been able to resist in place after place the march of multinational corporations across space. In Kentucky, KFTC lived up to its motto "Saving the Homeplace" and defeated "King Coal," which had been literally undermining their communities. ACNag prevented the nuclear industry and New York State from turning their place into a nuclear waste space. CCSC (Concerned Citizens of South Central) stopped the siting of a solid waste incinerator, thus saving their neighborhood from becoming a toilet for industrialism.

The ACNag bridge protest is a compelling example of the ability of environmental justice activists to interrupt the progress of industrialism. In blocking a road, the protesters are disrupting literally and symbolically a major artery of industrialism. Indeed, the restructuring of the economy, foreign policy, housing, and social practices around the needs of automotive transportation suggest that our society in the late 20th century could be termed a "car culture." In blocking this road, this lifeline of industrialism, the protesters confront the productive and symbolic capital of the culture and violate social norms.

In defense of their place ACNag activist embody their place, going so far as to identify themselves as Allegany County. The wearing of masks minimizes their individual indentities as they
assume the collective identity of the place where they are implaced. Their masks read "ALLE-
GANY" and "NO DUMP."

> From an early age, all children are inculcated with a necessary re-
> spect: pedestrians should always *give way* to automobiles. Such rules
> are mostly concerned with letting road users "go about their (and
> capitalism's) business," and are intended to coerce those who might
> obstruct their "rights of way" or infringe on their liberty. . . . For
> these reasons, if no other, the advent of recent road protests marks a
> radical challenge to the instrumental, one-dimensional, and codified
> ethos of the modern road. (Smith, 1997, p. 349, emphasis in original)

These activist grandparents in the road interrupt the industrialization
and homogenization of time and space and call to us to slow down and
consider this place. By forcing the industrial juggernaut to pause, if only
for a moment, their bodies give us pause. Such pause opens a space for
refuting the oft-repeated assertion of industrialism that progress is in-

evitable. In pausing, we stop the clock for a moment. In that moment, we can take the time to notice this particular place.

The names of these groups also highlight their implacement: Kentuckians for the Commonwealth, Allegany County Non-violent Action Group, Love Canal Homeowners Association, Warren County Citizens Concerned about PCBs, Concerned Neighbors in Action, Southwest Organizing Project, Mothers of East Los Angeles, Northeast Community Action Group.

The hegemonic discourse of industrialism is not a natural fact, but a rhetorical/political/social achievement that is perpetuated, but also challenged, through rhetorical/political/social practices. The sense of "naturalness" of industrialism has been achieved, in part, through trumpeting its successes in the rhetorics of progress and technological determinism, while muffling the cries of anguish and protest from its victims and resisters. In the 20th century, multiple antagonisms, from World Wars to genocides, from nuclear fallout to toxic pollution, from job displacement by automated technology to spiritual displacement by automated lives, have made clear the costs of industrialism and have created opportunities for voices of resistance to decry industrialism and articulate alternative visions of progress. Realizing such opportunities requires the construction of worlds, of contexts, wherein questioning industrialism makes sense.

In an industrial world the image events of radical environmental and environmental justice groups register as the antics of the unruly or, more harshly, the crimes of terrorists. More benignly, perhaps, the channel surfer sees them as just another spectacle to catch in a world of endless spectacles. The task of the critical rhetorician, I have suggested, is to construct new contexts, new worlds, that will enable a new "truth" about environmental tactics and activists. A truth that understands blocking a bulldozer to save a forest or community not as a crime, but as a heroic act of participatory democracy. A truth that understands calling such activists "Luddites" not as a means of dismissing them as anachronistic lunatics, but as a way of praising them as people who value community needs over technological dreams and profit motive greed. A truth that understands environmental activists and their rhetorical tactics not as isolated incidents, discordant notes during the march of industrial progress, but as part of a chorus of opposition over the centuries to the costs of industrialism, a chorus that constitutes a rich tradition of struggle that provides a spark of hope for those confronting a daunting future.

NOTES

CHAPTER I

1. The choice of the term "image event" is deliberate. Although an image event shares many of the characteristics of Daniel Boorstin's "pseudo-event" (1962), a term imported into rhetorical theory by David Berg (1972), I want to avoid the negative connotations of "pseudo." Plus, the ontological distinction Boorstin makes between reality and pseudo-events is untenable, which Boorstin acknowledges (1962, pp. 36–37). Katz and Dayan's "media event" (1986) implies too narrow a definition of media and too specialized and ritualistic an event for my purposes. John Fiske's conceptualization of "media event" (1994) explicitly challenges the ontological divide between reality and media events that trips up Boorstin. However, he also does not highlight the ocular aspects of media events to the extent I want to with the concept of image event. Although Guy Debord's concept of spectacle does highlight the visual quality (1983), it has negative connotations both generally and in Debord's work, where spectacles are associated with detachment from the directly lived, a pseudo-world, reciprocal alienation, the deceived gaze and false consciousness, the concrete inversion of life (1983, paragraphs 1, 2, 3, 8). Gronbeck's notion of the telespectacle (1995) deflects attention from images with the prefix "tele" and also suffers from negative connotations, which Gronbeck pleadingly attempts to overcome (1995, pp. 221–224, 234–235). My emphasis on images is justified by the qualitative differences between imagistic and linguistic discourses. For further reading on importance of different media to both formations of culture and patterns of perception and thinking, see Mumford (1934), Ong (1982), Havelock (1963), Innis (1951/1964, 1950/1972), McLuhan (1962, 1964), Carey (1989), Meyrowitz (1985), Postman (1985), Goody (1977), Gronbeck (1995), and Mander (1991), among others. For works that highlight images, see Berger (1972, 1980), Barthes (1981), Sontag (1977), and Dondis (1993), among others.

2. This claim is clearly true with respect to environmental groups, but is more controversial with regards to other groups working for social change. The key words in the sentence are "primary," "image events," and "mass media dissemination." Certainly, for centuries activists have performed civil disobedience and other acts in order to generate publicity and agitate for social change on various issues. Prominent examples would include the Boston Tea Party, Thoreau's refusal to pay a tax in protest of the Mexican War, and Gandhi's nonviolent campaign of resistance. Only with the advent of television, however, do conditions make possible a dramatic change in social protest tactics. Although civil rights activists, especially those under the leadership of Martin Luther King Jr., effectively staged image events, I would argue that image events were not their primary form of rhetorical activity. Instead, speeches, sermons, marches, boycotts, and small group interactions constituted the bulk of their rhetorical activity. Indeed, the only group that fits under my criteria prior to Greenpeace would be the Yippies of the late 1960s. Yippie leader Abbie Hoffman describes the new targets and tactics for the conditions in an age of television:

> America has more television sets than toilets. I began to understand those little picture tubes. If the means of production were the underpinnings of industrial society, then the means of communication served that function in a cybernetic world. And if labor was the essential ingredient for production, then information was that ingredient for mass communication. A modern revolutionary group heads for the television station, not the factory. It concentrates its energy on infiltrating and changing the image system. (quoted in Bowers, Ochs, and Jenson, 1971/1993, p. 21)

In Earth First! there is a clear Yippie connection. Mike Roselle, one of the founders of Earth First!, had been a Yippie activist (Foreman, 1991a, p. 216).

3. I do not correct or add to the sexist language used by some of those quoted in this work for the reason that such sexist language is not merely a matter of terminology but is often of theoretical or historical significance. For example, when theorists write on "man" as a universal category, often their analyses are only of men and pay little attention to the experience of women, though they proceed to universalize their findings.

4. For a fuller sense of the wide array of tactics and image events that Earth First! deploys, the 17 years of *Earth First!: The Radical Environmental Journal* are a good source.

5. For an account of this incident and other violent incidents against environmentalists, see Foreman (1991a, pp. 124–127), Helvarg (1994a, 1994b), and Rowell (1996).

6. My descriptions of the four sets of image events are drawn from a number of video sources. For the Greenpeace image event I viewed *Greenpeace's Greatest Hits* (Tatum, 1988) as well as CBS, NBC, and ABC News broadcasts of July 1, 1975. The Earth First! image events are culled primarily from network news broadcasts, especially the ABC News story "War in the Woods" (1987). Some images were broadcast in numerous stories. KFTC's own video, *Fighting for Justice*, was my source for their image event. I viewed the ACNag image event in the documentary *My Name is Allegany County* (O'Shaughnessy, 1993). Local

groups concerned with local issues tend to only get coverage from local news, the recordings of which are ephemeral.

7. CCHW has recently changed its name to the Center for Health, Environment, & Justice. Although this new name more accurately describes their mission and multiple activities, in this book I will stick with the initials CCHW. I do this for two reasons: the old name and initials have historical resonance and the group itself still uses them as well as the new name.

8. For a plethora of examples of environmental justice activity, see *Everyone's Backyard*, the newsletter of Citizens Clearinghouse for Hazardous Wastes.

9. This definition owes much to conversations with Barbara Biesecker.

10. I want to emphasize that my refusal to dismiss image events as the rude and crude antics of the unruly is neither a denial nor a celebration of their incivility. Rather, it is a protest against a double standard in incivility, whereby the incivility of the marginalized is an excuse to silence them while the incivility of the center is tacitly ignored or praised as "law and order." This is most stark in instances of civil disobedience, where civil protesters are often condemned as disorderly and arrested for disturbing the peace while police using excessive force to arrest protesters are praised and justified as enforcers of law and order, guardians against anarchy. The association of rhetoric with gentility from Aristotle to the present day not only makes marginal groups voiceless in the public sphere, it also tends to render them invisible to the academy. Such a disciplinary defining of incivility as outside the bounds of rhetoric leaves rhetorical theory unable to account for social change, blind to much of the activity of the public sphere, and incapable of understanding rhetoric.

11. As of November 1992, 69% of adults mentioned TV as where they get most of their news about what is happening in the world today. As of 1993, 46% of adults had a great deal of confidence in TV news, as compared to 31% in newspapers. See *TV Dimensions '95* (Papazian, 1995, pp. 240, 239).

12. Before the advent of image events, environmental change was primarily propelled by essays and books by the likes of Thoreau (1948/1966), Marsh (1864), Muir (1967/1968), Leopold (1949/1968), Carson (1962), and Abbey (1976), although landscape photography and art have had an unappreciated impact.

13. Besides Jamieson and Gronbeck, see Benson (1985, 1989), Brummett (1994), Condit (1990a), Lucaites (1997), Medhurst and Benson (1991), L. C. Olson (1991), and Tiemens (1989). Condit's focus on the visual images used by pro-life protesters is the most analogous to my study of the visual rhetoric of environmental groups.

14. Olson and Goodnight employ simplistic understandings of discursive and nondiscursive as words and images. For this discussion of their essay, I will work off of their understandings. In the rest of this book, however, I understand the discursive or discourse to include both the linguistic and the nonlinguistic. This issue will be taken up later.

15. For example, in an essay titled "Television Sound" (1987), Rick Altman argues that "a strong case can be made for the centrality of the sound track in the American commercial broadcast system and the other national systems that most resemble it" (p. 566). After attempting to make that case, Altman

concludes: "Only when the sound track reaches out to bring the two together [the TV flow and the household flow] do they fulfill their full mission" (p. 581). For Altman, then, the very telos of the televisual experience is only achieved through the intervention of sound.

16. Some argue that while time metaphors were characteristic of modernity, spatial metaphors are characteristic of postmodernism (see Harvey, 1989; Soja, 1989).

17. Olson and Goodnight's 1994 essay marks a significant radicalization of Goodnight's Habermassian conceptualization of the public sphere. Still, Habermas's ideal of the liberal public sphere exerts an idolic power over Goodnight's thought.

CHAPTER 2

1. The term "new social movement" is itself subject to debate. Cohen argues for "contemporary social movements" (1985, p. 667) and Melucci prefers "movement networks" or "movement areas" (1984, p. 828, 1985, p. 798). Since "new social movement" is the term with the widest currency at the moment, that is the term I shall use.

2. So named due to their focus on how the new social movements thematize issues concerning personal and collective identity.

3. Although agreeing with McGee's approach to social movement, I don't feel it necessary to dismiss the work of those in the sociological tradition as worthless. Studies of the rhetoric of particular groups that call themselves or are called by others "social movements" are often insightful. For example, Simons' positive valuation of bottom-up groups and his validation of coercion as a tactic of persuasion are important theoretical and practical moves.

4. Further evidence of Simons' theoretical inconsistency: although Simons always studied social movements as if they could be reduced to particular organizations, in two important essays he did not define them that way. Instead, he defines social movements as "struggles" (1991, p. 100) or "sustained *efforts* by noninstitutionalized collectivities to mobilize resources, resist counterpressures, and exert external influence in behalf of a cause" (Simons et al., 1984, p. 794). Such a definition is more akin to *Webster's* second definition and is more amenable to rhetoric because it suggests acts (often rhetorical) and not organizations as the locus of study.

5. It is interesting to note that while in his 1970 essay Simons stressed the importance of theory in illuminating rhetorical requirements, strategies, and tactics and in suggesting parameters and directions, as critics chipped away at his theoretical edifice Simons started to value theory less, wondering why critics were "so exercised" over theoretical questions (1980, p. 309) and regretting "abstract and seemingly interminable discussions" that he now saw as "pseudodisputes" that weigh rhetoricians down (1991, p. 94).

6. Although McGee's point is well taken, it is overstated. I will dwell on the importance of history for the practice of critical rhetoric in Chapter 7.

7. Jonathon Porritt, in *Seeing Green*, develops this point:

> Both are dedicated to industrial growth, to the expansion of the means of pro-
> duction, to a materialist ethic as the best means of meeting people's needs, and
> to unimpeded technological development. Both rely on increasing centraliza-
> tion and large-scale bureaucratic control and coordination. From a viewpoint
> of narrow scientific rationalism, both insist that the planet is there to be con-
> quered, that big is self-evidently beautiful, and that what cannot be measured is
> of no importance. Economics dominates; art, morals, and social values are all
> relegated to a dependent status. . . . They are united in one, all-embracing "su-
> per-ideology," which, for the sake of convenience, I intend to call industrialism
> (1984, pp. 43–44).

8. This notion of competing discourses is akin to McGee's brief description of a social field consisting of competing political myths that function to redefine reality and provide social unity and collective identity (1975, pp.
246–247).

9. Harold Schlechtweg makes this point in his excellent essay on the relation of rhetoric and philosophy in a postmodern age (Schlechtweg, 1990).

CHAPTER 3

1. Although there is some disagreement over the importance or even the
existence of progress in earlier historical periods, there is a general consensus
that progress achieves dominance in the 18th and 19th centuries. This is evident
in the formulation of the modern notion of inevitable progress in the philosophies of history of the time by such figures as Voltaire, Turgot, Condorcet,
Hegel, Marx, Proudhorn, and Comte. For accounts of the rise and importance
of progress, see Nisbet (1980), Bury (1932/1955), Blumenberg (1985), Lasch
(1991), Collingwood (1956), Capra (1983), Paepke (1993), Stent (1978), Merchant (1992), Oelschlaeger (1991), and Shiva (1988), among others.

2. Many theorists have explored this connection. The Frankfurt School,
which understood the domination of nature as being at the heart of the reversal
of the Enlightenment, argued that the domination of nature involved not only
the domination of external nature but the domination of other humans and the
domination of one's own inner nature. Although this domination of humans as
nature is most readily seen in colonizing practices, it is also apparent in such an
esteemed practice as medicine, where the practice of experimenting on and tor-
turing animals quite easily slides into the practice of experimenting on and tor-
turing humans.

In a similar vein, John Berger argues that the reduction of the animal pre-
figured the process by which humans were reduced to isolated productive and
consuming units: "Nearly all modern techniques of social conditioning were
first established with animal experiments" (1980, p.11).

Many feminists, including Sherry Ortner (1974), Susan Griffin (1978), Car-
olyn Merchant (1991, 1992), Donna Haraway (1991), and Judith Butler (1990),

have explored the political and social effects of the imposition of certain concepts of nature and what's natural upon women.

Perhaps Langdon Winner put it best: "Nature will justify anything. Its text contains opportunities for myriad interpretations" (1986, p. 137). Others who discuss the nature–culture connection include Collingwood (1945), Evernden (1992), Glacken (1967), Nash (1967/1973), and Snyder (1990).

3. For accounts of the establishment of this hegemonic meaning of nature, see Berman (1984), Capra (1983), Collingwood (1945), Evernden (1992), Heidegger (1977), Leiss (1972), Oelschlaeger (1991), Toulmin (1990), and White (1968).

4. Although this collection of examples is fairly eclectic, I think it points to the pervasiveness of progress in all social spheres: economic, environmental, industrial, athletic, and technological.

CHAPTER 4

1. The critics of postmodernism are legion and include such disparate groups as the proponents of liberalism, Marxism, and critical cultural studies. Names would include Habermas (1987), Hall (1986), Eagleton (1991), Norris (1992), and Geras (1987).

2. I recognize that I am writing in a modern form while arguing for postmodern claims. There are many reasons for this, not the least of which is that I am precariously perched in a modern institution, the university. McKerrow addresses some of these problems (1991a).

3. Similar characterizations of modernism as being essentially about progress/industrialism can be found in the work of a number of theorists and historians. David Harvey identifies modernism "with the belief in linear progress, absolute truths, the rational planning of ideal social orders, and the standardization of knowledge and production" (1989, p. 9) and relies on Habermas to define the project of modernity as

> an extraordinary intellectual effort on the part of Enlightenment thinkers "to develop objective science, universal morality and law, and autonomous art according to their inner logic." The idea was to use the accumulation of knowledge generated by many individuals working freely and creatively for the pursuit of human emancipation and the enrichment of daily life. The scientific domination of nature promised freedom from scarcity, want, and the arbitrariness of natural calamity. The development of rational forms of social organization and rational modes of thought promised liberation from the irrationalities of myth, religion, superstition, release from the arbitrary use of power as well as from the dark side of our own human natures. Only through such a project could the universal, eternal, and the immutable qualities of all of humanity be revealed. (1989, p. 12)

William Leiss describes the project of modernity in this manner: "Through a collective social enterprise, extending over many generations and paced by the march of science and technology, the human species would fulfill its destiny by gaining complete control over the forces of the natural world, appropriating to the full its resources for the satisfaction of human needs" (1972, p. xi). As Ian Angus simply puts it, "Modernity can be defined as the domination of nature in order to produce a wealth of commodities which are intended to sustain a community of mutually recognizing free and equal subjects" (1989a, p. 96; see also Blumenberg, 1985; Capra, 1983; Oelschlaeger, 1991; Rifkin, 1980; Sale, 1995; and Shiva, 1988).

4. This characterization of postmodernism is also indebted to more general sources not cited, in particular the work of Jacques Derrida and Jean-Francois Lyotard, and benefitted from conversations with Barbara Biesecker. Of the many general descriptions of postmodernism, particularly cogent are the accounts by Harvey (1989) and White (1991).

5. This concern is expressed by the contributors to *Reinventing Nature? Responses to Postmodern Deconstruction,* Soule, and Lease Eds. (1995).

6. Although not all versions of ecofeminism espouse essentialism, I am explicitly referring to certain radical versions that are not incidental to ecofeminism, but rather are central, if not dominant. I would include among ecofeminists espousing essentialist versions Mary Daly, Susan Griffin, Ynestra King, Marti Kheel, and Gloria Orenstein. Though I do not have the space to make the argument here, see Stabile (1994). Many, most notably Gayatri Spivak, suggest that at times essentializing is a useful tactic. My position, argued in the next few pages, is that essentializing is a dangerous and counterproductive tactic that reproduces the oppressive hierarchies of modernism. That said, I do not wish to bash ecofeminism, which I recognize as a heterogenous ideology that often problematizes nature in politically important ways.

7. I am using the term "ideograph" (McGee, 1980a) here instead of "concept" or the more postmodern "sign" because I think it more effectively points to the constructed character of "Nature" and the rhetorical and political aspects of that construction.

8. Originally, they were nine: National Wildlife Federation, Sierra Club, National Audubon Society, Wilderness Society, Izaak Walton League, Natural Resources Defense Council, Environmental Defense Fund, Environmental Policy Center, and Friends of the Earth (Gottlieb, 1993, p. 118).

9. "Nazi forestry" was the Wilderness Society advisor's term. I include it here to point to the recognition of the Wilderness Society that Plum Creek Timber is among the worst timber companies due to its destructive practices. However, the term also raises the serious question of whether "Nazi" gets bandied about too freely, thus diluting its meaning and belittling the horror of the Holocaust. I do not have an answer, but wanted to highlight the issue.

10. Their impact is most dramatically illustrated in the case of strip mining. In a state where coal is king, after being rebuffed by the governor and the state supreme court, KFTC succeeded in getting passed a constitutional amendment banning broad form deed strip mining. Despite being outspent by the coal industry three-to-one, KFTC's amendment won 82.5% of the vote. For more examples, see Zuercher (1991).

11. The Love Canal Homeowners Association's success at Love Canal had spurred Congress to establish the original Superfund (Gottlieb, 1993, p. 188).

12. On financial grounds alone, this is a pivotal shift, for within the parameters of conventional politics the strategies of lobbying, litigating, and scientific expertise require significant financial resources. The mainstream groups' millions are no match for industry's billions. Even Sierra Club political director Daniel Weiss admits, "Buying influence is an arms race that the environmental community cannot win" (quoted in Dowie, 1995, p. 86).

13. There are numerous examples of corporations giving in to activists' demands in order to protect their images. Burger King stopped using cattle from tropical rainforest land. StarKist/Heinz agreed to stop slaughtering dolphins when catching tuna. Various clothing retailers have pledged to avoid buying from sweatshops. Disney abandoned plans for Disney's America on a Civil War battlefield in Virginia.

CHAPTER 5

1. For some examples, see Bagdikian (1987); Enzensberger (1974); Helvarg (1994a); Herman and Chomsky (1988); Horkheimer and Adorno (1972); Jhally (1989); Kline (1989); Parenti (1993); Schiller (1989); and Smythe (1981).

2. CBS is chosen because at the time, under the regime of Walter Cronkite, it was the broadcast of record. Not only did CBS News have the largest audience at the time, but as Hogan notes in studying the nuclear freeze movement, "as comparative studies of news content have shown, the differences across networks are largely cosmetic; all appear governed by the same values" (1994, pp. 142–143).

3. For examples and extended discussion of this dilemma, see Lange (1990); Branham and Pearce (1985); Marcuse (1964); Johnstone (1974); and Peterson (1988).

4. Although environmental activists have utilized the science of ecology, it is important to recall the title of a famous text that names ecology "the subversive science" (Shepherd & McKinley, 1969).

CHAPTER 6

1. Much of this research has been conducted in the area of interpersonal studies. See Archer and Akert (1977); Argyle, Alkema, and Gilmour (1971); Berman, Shulman, and Marwit (1976); Bugental, Koswan, Loue, and Fox (1970); Burgoon, Buller, and Wordell (1989); McMahan (1976); Mehrabian (1967); Posner, Nissen, and Klein (1976). I acknowledge that this is not a strictly analogous situation and that the relative force of images and words is still an open debate. However, I think images have something to do with why television has displaced radio to such an extent. That said, I will grant that in different genres and in different forms of television around the world, the ratio of importance between images and words varies.

2. More recent military actions reinforce this lesson of the Vietnam War.

The United States orchestrated the Gulf War in a manner that produced the impression of a virtual war resembling a video game. While the slaughter of Iraqis was acceptable, images of the carnage were banished from the electronic public sphere. In Somalia, the image of one dead soldier being dragged through the streets routed U.S. forces. For extended discussions of these examples, see Donovan and Scherer (1992), Orwin (1996), and Sharkey (1993).

3. All of the following quotations related to the strategy of Reagan's media team are taken from the Bill Moyers video "Illusions of News," which was part of his series *The Public Mind* (1989).

4. "Atrocity drive" is the term Mark Crispin Miller uses in his manuscript *Spectacle* (n.d.) to describe the media propaganda campaign against Iraq during the Gulf War. Atrocity drives are characterized by the systematic propagation of falsehoods and half-truths in order to demonize one's opponents. It is more sophisticated than simple name-calling and involves the strategic use of lies and exaggerations to create a context that calls for and justifies one's objectives. I think the term aptly describes the efforts of anti-environmental forces (corporations, conservative congressional members, Rush Limbaugh and other right-wing talkshow hosts) as they rely on isolated incidents, questionable anecdotes, and outright fabrications to justify dismantling environmental regulations protecting the air, water, people, and endangered species from the consequences of industrialism.

5. Over 80% of the television news audience is white. Some 78% of the audience earns more than $20,000 per year, 60% of the audience earns over $30,000. See *TV Dimensions '95*, edited by Ed Papazian (p. 1995, p. 244).

6. That other models are too complicated is a weak excuse. If we cannot get beyond common sense with our students, then what is the point of teaching?

7. Members would include Innis (1950/1972, 1951/1964); McLuhan (1962, 1964); Carey (1989); Baudrillard (1972/1981, 1983); and Meyrowitz (1985), among others.

8. Although Baudrillard is the example here, many media theorists also share the metaphysical commitments of the transmission model. Carey's ritual model readily springs to mind.

9. I chose to keep the citations from a manuscript that was later published (Peters, 1994) because the exact wording better fit my purposes, though I think the revisions in the published version are getting at the same points. Judge for yourself. The p. 11 quote now reads: "Mass communication . . . is in fact an old form, maybe the most basic form of communication. One turn of conversation awaits and calls for another turn; in mass communication, the single turn is displayed in all its alarming solitude" (pp. 132–133). The p. 12 quote now reads: "Interpersonal communication could be seen as series of interlocked acts of mass communication. Conversation, to put it a bit archly, could be seen as two or more people taking turns 'broadcasting' at each other" (p. 132). The p. 13 quote now reads: "the radical indeterminacies of effects and reception" (p. 131).

10. Indeed, Spivak suggests that Derrida comes perilously close to making an argument for historical necessity (1976, pp. lxxx–lxxxi).

11. Derrida justifies retaining the old name due to those very connotations

and the necessity of placing his analysis within the already constituted historical field (1982, pp. 329–330; 1974/1976, p. 56).

12. A difficult task, but not a new one. Some excerpts from David Hume's *A Treatise of Human Nature* shows that even under the shadow of the Cartesian self, the subject has remained a contested category:

> There are some philosophers who imagine we are every moment intimately conscious of what we call our *self*; that we feel its existence and its continuance in existence; and are certain, beyond the evidence of a demonstration, both of its perfect identity and simplicity. . . . For my part, when I enter most intimately into what I call *myself*, I always stumble on some particular perception or other, of heat or cold, light or shade, love or hatred, pain or pleasure. I never can catch *myself* at any time without a perception, and never can observe anything but the perception. . . . I may venture to affirm of the rest of mankind, that they are nothing but a bundle or collection of different perceptions, which succeed each other with an inconceivable rapidity, and are in a perpetual flux and movement. . . . What then gives us so great a propension to ascribe an identity to these successive perceptions, and to suppose ourselves possessed of an invariable and uninterrupted existence through the whole course of our lives? . . . We feign the continued existence of the perceptions of our senses, to remove the interruption; and run into the notion of a *soul*, and *self*, and *substance*, to disguise the variation. (1739/1962, pp. 300–304, emphasis in original)

I have indulged in such an extensive quotation for two reasons. First, Hume highlights issues that are central to current postmodern debates about the subject. Note how similar Hume's "bundle or collection of different perceptions" is to Chang's "bundles of practices." Many other thinkers over the centuries have dealt with these issues, which points to the contestation of the subject as a continuous conversation. A nice example of this conversation is the exchange between Edmund Burke, with his nonessentialized, geographically and historically localized sense of the subject, and Thomas Paine, who argued for inherent, universal human rights. Second, given the breadth and depth of the conversation, my excursion is meant to be tentative, not definitive.

13. Gaonkar makes a similar critique of the close readings in the area of the rhetoric of science (1993, 1997).

14. See also Campbell (1990a, p. 349). For evidence see Leff and Sach's endnote #23 (1990, p. 271).

15. Although Leff uses the terms "work" and "text" interchangeably, in poststructuralist thought the differences are significant. For the most concise discussion of the differences, see Barthes (1986, pp. 49–64).

16. Note how the subjectivity of the critic escapes scrutiny.

17. The similarity of Barthes's and McGee's descriptions of the text resonates. According to Barth, "If the text is subject to some form, this form is not unitary, architectonic, finite: it is the fragment, the shards, the broken or obliterated network—all the movements and inflections of a vast 'dissolve,' which permits both overlapping and loss of messages" (1974/1987, p. 20). According to McGee, "Texts are understood to be larger than the apparently finished dis-

course that presents itself as transparent. The apparently finished discourse is in fact a dense reconstruction of all the bits of other discourses from which it was made. It is fashioned from what we can call 'fragments'" (1990, p. 279).

18. For examples, see Lewis (1985, 1991); and Morley (1980, 1981).

19. Note that McGee's story assumes a Foucauldian shift. I prefer a Derridean perspective that sees a general unfolding of the always already existent structural conditions that make possible signs, texts, subjects, audiences, and meanings.

20. It is important to note that my critique here is not of all of McGee's work, but rather is focused on the essay in the *Western Journal of Speech Communication*. Such a focus is justifiable on two grounds. First, that special issue of *Western* purposefully divided the field of rhetorical criticism into two alternatives, with McGee's programmatic essay being one of them. Both the issue and the essay fulfilled their purposes, with McGee's essay gaining deservedly wide currency. Second, the essay exhibits a tension between modernism and postmodernism characteristic of much of McGee's work, so I do not think that what I point to is an anomaly. That said, such inconsistency is not a sin but, rather, a sign of daring and openness. Further, as Derrida notes, to a certain extent such is the unavoidable fate of writing (1976, pp. 157–164; 1981), "that, in a certain sense, it is impossible 'not to deconstruct/be deconstructed'" (Spivak, 1976, p. lxxviii). This is an important essay and my critique is both an acknowledgment of that and an attempt to purge McGee's argument of metaphysical baggage that blunts the radicalness of his insights.

21. McGee's invocation of "the world" is suspect. What is in the real world and what is outside of it? Surely, there is nothing outside of the world. Conversely: "*There is nothing outside of the text*" (Derrida, 1976, p. 158, emphasis in original).

22. Although this is not the place for a deconstruction of audience research, I want to mention that the very form of questioning presumes a transparent, self-aware subject.

CHAPTER 7

1. "Postmodern" is a notoriously vague and slippery term that is also unavoidable. I think it is best to think of it as what Raymond Williams terms a structure of feeling (1977b, pp. 128–135). To reiterate my definition from Chapter 4, postmodernism can be characterized in part by the conjunction of the following elements: a decentering of the subject as origin, end, and arbiter of theory and practice; a destabilization or fragmentation of all kinds of identity; a lack of belief in any foundation, totality, transcendental signified, or grand narrative; recognition of a plurality of discourses in an open social field that cannot be subsumed under any definitive organizing principle (Angus, 1993); a shift from a thematics of need and truth to desire and power; a generalized

awareness of limits, particularly the limits of reason (Laclau, 1990, p. 3); a change in material conditions, including the disappearance of Nature as the great referent that ontologically grounds Western epistemology; and the rise of both image and micropolitics.

2. For an explicit statement of this position, see Laclau and Mouffe (1985, pp. 107–112).

REFERENCES

Abbey, E. (1976). *The monkey wrench gang.* New York: Avon Books.

ABC News. (1975, July 1). New York: ABC.

ABC News. (1987, August 10). *War in the woods.* New York: ABC.

ABC News. (1993, August 19). *America agenda.* New York: ABC.

Aden, R. C. (1994). The enthymeme as postmodern argument: Condensed, mediated argument than and now. *Argumentation and Advocacy, 31,* 54–63.

Adorno, T., & Horkheimer, M. (1977). The culture industry: Enlightenment as mass deception. In J. Curran, M. Gurevitch, & J. Woolacott (Eds.), *Mass communication and society* (pp. 349–383). London: Edward Arnold.

Agee, W. K., Ault, P. H., & Emery, E. (1985). *Introduction to mass communications.* New York: Harper & Row.

Altman, R. (1987). Television sound. In H. Newcomb (Ed.), *Television: The critical view* (pp. 566–584). Oxford, UK: Oxford University Press.

Angus, I. (1984). *Technique and enlightenment: Limits of instrumental reason.* Washington, DC: Centre for Advanced Research in Phenomenology and University Press of America.

Angus, I. (1989a). Circumscribing postmodern culture. In I. Angus & S. Jhally (Eds.), *Cultural politics in contemporary America* (pp. 96–107). New York: Routledge.

Angus, I. (1989b). Media beyond representation. In I. Angus & S. Jhally (Eds.), *Cultural politics in contemporary America* (pp. 333–346). New York: Routledge.

Angus, I. (1992). The politics of common sense: Articulation theory and critical communication studies. In S. Deetz (Ed.), *Communication yearbook, 15* (pp. 535–). Newbury Park: Sage.

Angus, I. (1993). Learning to stop: A critique of general rhetoric. In I. Angus & L. Langsdorf (Eds.), *The critical turn: Rhetoric and philosophy in postmodern discourse* (pp. 175–211). Carbondale: Southern Illinois University Press.

Angus, I., & Jhally, S. (Eds.). (1989a). *Cultural politics in contemporary America.* New York: Routledge.

Angus, I., & Jhally, S. (1989b). Introduction. In I. Angus & S. Jhally (Eds.), *Cultural politics in contemporary America* (pp. 1–16). New York: Routledge.

Angus, I. H., & Lannamann, J. W. (1988). Questioning the institutional boundaries of U.S. communication research: An epistemological inquiry. *Journal of Communication 38:3* (Summer), 62–74.

Archer, D., & Akert, R. M. (1977). Words and everything else: Verbal and nonverbal cues in social interpretation. *Journal of Personality and Social Psychology, 35*, 443–449.

Argyle, M., Alkema, F., & Gilmour, R. (1971). The communication of friendly and hostile attitudes by verbal and nonverbal signals. *European Journal of Social Psychology, 1*, 385–402.

Aristotle. (1962/1981). *Nicomachean ethics*. Indianapolis: Bobbs-Merrill Educational Publishing.

Aristotle. (1991). *On rhetoric: A theory of civic discourse*. New York: Oxford University Press.

Arnold, C. C., & Bowers, J. W. (Eds.). (1984). *Handbook of rhetorical and communication theory*. Boston: Allyn & Bacon.

Aune, J. A. (1983). Beyond deconstruction: The symbol and social reality. *Southern Speech Communication Journal, 48*(Spring), 255–268.

Bagdikian, B. (1987). *The media monopoly*. Boston: Beacon Press.

Balthrop, V. W. (1989). W(h)ither the public sphere? An optimistic reading. In B. E. Gronbeck (Ed.), *Spheres of argument: Proceedings of the sixth SCA/AFA conference on argumentation* (pp. 20–25). Annandale, VA: Speech Communication Association.

Barthes, R. (1972). *Mythologies*. New York: Hill & Wang.

Barthes, R. (1974). *S/Z*. New York: Hill & Wang.

Barthes, R. (1977). *Image, music, text*. New York: Hill & Wang.

Barthes, R. (1981). *Camera lucida*. New York: Noonday Press.

Barthes, R. (1986). *The rustle of language*. New York: Hill & Wang.

Bateson, G. (1972). *Steps to an ecology of mind*. New York: Ballantine.

Baudrillard, J. (1972/1981). *For a critique of the political economy of the sign*. New York: Telos.

Baudrillard, J. (1983). New York: Semiotext(e).

Bazan, G. (1996, March 17). First, embrace new attitude about growth. *Centre Daily Times*, p. 6A.

Beasley, C. (1991). Moore takes on all. *Buzzworm, 3*(3), 38–43.

Benjamin, W. (1968a). Theses on the philosophy of history. In H. Arendt (Ed.), *Illuminations* (pp. 253–264). New York: Schocken Books.

Benjamin, W. (1968b). The work of art in the age of mechanical reproduction. In H. Arendt (Ed.), *Illuminations* (pp. 217–252). New York: Schocken Books.

Benson, T. (Ed.). (1985). *Speech communication in the 20th century*. Carbondale: Southern Illinois University Press.

Benson, T. (1989). *Reality fictions: The films of Frederick Wiseman*. Carbondale: Southern Illinois University Press.

Berg, D. M. (1972). Rhetoric, reality, and mass media. *Quarterly Journal of Speech, 58*(October), 255–263.

Berger, J. (1972). *Ways of seeing.* Middlesex, UK: Penguin Books.

Berger, J. (1980). *About looking.* New York: Pantheon Books.

Berger, J. (1984). *And our faces, my heart, brief as photos.* London: Writers & Readers.

Berlin, J. A. (1990). Postmodernism, politics, and histories of rhetoric. *Pre/Text, 11,* 170–187.

Berlo, D. K. (1960). *The process of communication.* New York: Holt, Rinehart & Winston.

Berman, H. J., Shulman, A. D., & Marwit, S. J. (1976). Comparison of multidimensional decoding of affect from audio, video, and audiovideo recordings. *Sociometry, 39,* 83–89.

Berman, M. (1984). *The reenchantment of the world.* Toronto: Bantam Books.

Berry, W. (1987). *Home economics.* San Francisco: North Point Press.

Biesecker, B. (1989a). Recalculating the relation of the public and technical spheres. In B. E. Gronbeck (Ed.), *Spheres of argument: Proceedings of the sixth SCA/AFA conference on argumentation* (pp. 66–70). Annandale, VA: Speech Communication Association.

Biesecker, B. (1989b). Rethinking the rhetorical situation from within the thematic of difference. *Philosophy and Rhetoric, 22,* 110–130.

Biesecker, B. (1992). Michel Foucault and the question of rhetoric. *Philosophy and Rhetoric, 25,* 351–364.

Birdsell, D. S. (1989). Critics and technocrats. In B. E. Gronbeck (Ed.), *Spheres of argument: Proceedings of the sixth SCA/AFA conference on argumentation.* Annandale, VA: Speech Communication Associations.

Birdwhistell, R. L. (1955). Background to kinesics. *ETC, 13,* 10–18.

Birdwhistell, R. L. (1968). Communication. In D. L. Sills (Ed.), *International encyclopedia of social sciences* (Vol. 3, pp.). New York: Macmillan and Free Press.

Blumenberg, H. (1985). *The legitimacy of the modern age.* Cambridge, MA: MIT Press.

Boorstin, D. J. (1962). *The image or what happened to the American dream.* New York: Atheneum.

Booth, W. (1998, September, 1998). Logging protester killed by falling redwood tree. *Washington Post,* p. A2.

Borgmann, A. (1992). The postmodern economy. In S. Cutcliffe, S. Goldman, M. Medina, & J. Sanmartin (Eds.), *New worlds, new technologies, new issues* (pp. 000). Bethleham, PA: Lehigh University Press.

Bowers, J. W., & Ochs, D. J. (1971). *The rhetoric of agitation and control.* Prospect Heights, IL: Waveland Press.

Bowers, J. W., Ochs, D. J., & Jensen, R. J. (1993). *The rhetoric of agitation and control* (2nd rev. ed.). Prospect Heights, IL: Waveland Press.

Branham, R. J., & Pearce, W. B. (1985). Between text and context: Toward a rhetoric of contextual reconstruction. *Quarterly Journal of Speech, 71,* 19–36.

Breathed, B. (1990). *Happy trails.* Boston: Little, Brown.

Brock, B. L., Scott, R. L., & Chesebro, J. W. (1989). *Methods of rhetorical criticism: A twentieth-century perspective.* Detroit, IN: Wayne State University Press.

Brown, M., & May, J. (1991). *The Greenpeace story.* New York: Dorling Kindersly.

Browne, S. (1988). Edmund Burke's "Letter to a noble lord": A textual study in political philosophy and rhetorical action. *Communication Monographs, 55,* 215–229.

Browne, S. (1993). Reading public memory in Daniel Webster's "Plymouth Rock oration." *Western Journal of Communication, 57,* 464–477.

Browne, S. (1997). Webster's eulogy and the tropes of public memory. In T. Benson (Ed.), *Rhetoric and political culture in nineteenth-century America* (pp. 39–46). East Lansing, MI: Michigan State University Press.

Brummett, B. (1994). *Rhetoric in popular culture.* New York: St. Martin's Press.

Bugental, D. E., Kaswan, J. W., Love, L. R., & Fox, M. N. (1970). Child versus adult perception of evaluative messages in verbal, vocal, and visual channels. *Developmental Psychology* (Vol. 2, pp. 367–375).

Burgess, J. (1993, October 21). Communications chiefs urge deregulation. *Washington Post,* pp. D1, D10.

Burgoon, J. K., Buller, D. B., & Woodell, W. G. (1989). *Interpersonal communication.* New York: Harper & Row.

Burgoon, M., Hunsaker, F. G., & Dawson, E. J. (1994). *Human communication.* Thousand Oaks, CA: Sage.

Bury, J. (1932/1955). *The idea of progress: An inquiry into its growth and origin.* New York: Dover.

Butler, J. (1990). *Gender trouble: Feminism and the subversion of identity.* New York: Routledge.

Butler, J. (1992). Contingent foundations: Feminism and the question of "postmodernism." In J. Butler & J. Scott (Eds.), *Feminists theorize the political* (pp. 3–21). New York: Routledge.

Calhoun, C. (Ed.). (1992). *Habermas and the public sphere.* Cambridge, MA: MIT Press.

Calinescu, M. (1987). *Five faces of modernity.* Durham, NC: Duke University Press.

Campaigning for environmental justice. (1993). *Everyone's Backyard, 11*(1), 6–7.

Campbell, J. A. (1975). The polemical Mr. Darwin. *Quarterly Journal of Speech, 61,* 375–390.

Campbell, J. A. (1986). Scientific revolution and the grammar of culture: The case of Darwin's *Origin. Quarterly Journal of Speech, 72,* 351–376.

Campbell, J. A. (1987). Charles Darwin: Rhetorician of science. In J. S. Nelson, A. Megill, & D. N. McCloskey (Eds.), *The rhetoric of the human sciences* (pp. 69–86). Madison: University of Wisconsin Press.

Campbell, J. A. (1990a). Between the fragment and the icon: Prospect for a rhetorical house of the middle way. *Western Journal of Speech Communication, 54*(3), 346–376.

Campbell, J. A. (1990b). Darwin, Thales, and the milkmaid. In R. Trapp & J. Schuetz (Eds.), *Perspectives on argumentation* (pp. 207–220). Prospect Heights, IL: Waveland Press.

Campbell, J. A. (1990c). Scientific discovery and rhetorical invention: Darwin's path to natural selection. In H. W. Simons (Ed.), *The rhetorical turn* (pp. 58–89). Chicago: University of Chicago Press.

Campbell, J. A. (1990d). *On the way to the "Origin": Darwin's evolutionary insight and its rhetorical transformation* (the Van Zelst lecture in communication). Evanston, IL: Northwestern University School of Speech.

Campbell, K. K. (1971). The rhetoric of black nationalism: A case study in self-conscious criticism. *Central States Speech Journal, 22,* 151–160.

Capra, F. (1983). *The turning point: Science, society, and the rising culture.* New York: Bantam Books.

Carey, J. W. (1986). Why and how? The dark continent of American journalism. In R. Manoff & M. Schudson (Eds.), *Reading the news* (pp. 146–196). New York: Pantheon Books.

Carey, J. W. (1989). *Communication as culture.* Winchester, MA: Unwin Hyman.

Carlson, A. C. (1986). Gandhi and the comic frame: "Ad bellum purificandum." *Quarterly Journal of Speech, 72,* 446–455.

Casey, E. S. (1997). *The fate of place: A philosophical inquiry.* Berkeley and Los Angeles: University of California Press.

Cathcart, R. S. (1972). New approaches to the study of movements: Defining movements rhetorically. *Western Speech, 36*(Spring), 82–88.

Cathcart, R. S. (1978). Movements: Confrontation as rhetorical form. *Southern Speech Communication Journal, 43*(Spring), 233–247.

Cathcart, R. S. (1980). Defining social movements by their rhetorical form. *Central States Speech Journal, 31*(Winter), 267–273.

Cathcart, R. S. (1983). A confrontation perspective on the study of social movements. *Central States Speech Journal, 34*(Spring), 69–73.

CBS News. (1975, July 1). New York: CBS.

CBS News. (1990, May 13). New York: CBS.

Central States Speech Journal. (1980, Winter). *31,* 306–315.

Chang, B. G. (1985). The eclipse of being: Heidegger and Derrida. *International Philosophical Quarterly, 25,* 113–138.

Chang, B. G. (1986). Communication after deconstruction: Toward a phenomenological ontology of communication. *Studies in Symbolic Interaction, 7A,* 13–32.

Chang, B. G. (1987). Deconstructing the audience: Who are they and what do we know about them? *Communication Yearbook, 10,* 649–665.

Chang, B. G. (1996). *Deconstructing communiction: Representation, subject, and economies of exchange.* Minneapolis: University of Minnesota Press.

Charland, M. (1991). Finding a horizon and telos: The challenge to critical rhetoric. *Quarterly Journal of Speech, 77,* 71–74.

Charland, M. (1997). Anxious oratory—Anxious criticism. In T. Benson (Ed.), *Rhetoric and political culture in nineteenth century America* (pp. 157–162). East Lansing, MI: Michigan State University Press.

Charland, M., & Lucaites, J. L. (1989). The legacy of <liberty>: Rhetoric, ideology and aesthetics in the postmodern condition. *Canadian Journal of Social and Political Theory, 13,* 31–48.

Chase, S. (Ed.). (1991). *Defending the earth: A dialogue between Murray Bookchin and Dave Foremen.* Boston: South End Press.

Chasnoff, D. (Producer). (1991). *Deadly deception: General Electric, nuclear weapons and our environment* [Film]. Women's Educational Media.

Chavis-Legerton, R. (1993). Youth in action. *Southern Exposure, 21*(4), 11.

Clark, N. (1996). The critical servant: An Isocratean contribution to critical rhetoric. *Quarterly Journal of Speech, 82,* 111–124.

Clinton, B. (1996, August 29). In his own words. *New York Times,* pp. B11.

Cloud, D. L. (1994). The materiality of discourse as oxymoron: A challenge to critical rhetoric. *Western Journal of Communication, 58*(Summer), 141–163.

Cockburn, A. (1993, November 1). Beat the devil. *Nation,* pp. 486–487.

Cockburn, A. (1995a, February 6). Beat the devil. *Nation,* 156–157.

Cockburn, A. (1995b, February 20). Beat the devil. *Nation,* p. 228.

Cockburn, A. (1995c, March 6). Beat the devil. *Nation,* pp. 299–300.

Cockburn, A. (1995d, April 3). Beat the devil. *Nation,* pp. 443–444.

Cockburn, A. (1995e, May 1). Beat the devil. *Nation,* pp. 588–599.

Cockburn, A. (1996, April 8). Beat the devil. *Nation,* pp. 9–10.

Cockburn, A. (1997, February 10). Beat the devil. *Nation,* pp. 9–10.

Cockburn, A., & St. Clair, J. (1994, December 19). After Armagedden: Death and life for America's Greens. *Nation,* pp. 760–765.

Cockburn, A., & St. Clair, J. (1995, April 24). Stop him before he cuts again: Wilderness chief in tree massacre. *Nation,* pp. 556–559.

Cohen, J. (1998, June 8). Stories TV doesn't tell. *Nation,* pp. 7, 38.

Cohen, J. L. (1985). Strategy or identity: New theoretical paradigms and contemporary social movements. *Social Research, 52,* 663–716.

Collingwood, R. G. (1945). *The idea of nature.* Oxford, UK: Clarendon Press.

Collingwood, R. G. (1956). *The id¹ of history.* New York: Oxford University Press.

Commoner, B. (1993). Past and future of the grassroots. In *Ten years of triumph* (pp. 24–25). United States: Citizen's Clearinghouse for Hazardous Wastes.

Communication Studies. (1991, Spring). *42,* 94–101.

Condit, C. M. (1990a). *Decoding abortion rhetoric: Communicating social change.* Urbana: University of Illinois Press.

Condit, C. M. (1990b). Rhetorical criticism and audiences: The extremes of McGee and Leff. *Western Journal of Speech Communication, 54*(Summer), 330–345.

Condit, C. M. (1994). Hegemony in a mass-mediated society: Concordance about reproductive technologies. *Critical Studies in Mass Communication, 11,* 205–230.

Condit, C. M., & Lucaites, J. L. (1991). The rhetoric of equality and the expatriation of African-Americans, 1776–1826. *Communication Studies, 42,* 1–21.

Condit, C. M., & Lucaites, J. L. (1993). *Crafting equality: America's Anglo-African word.* Chicago: University of Chicago Press.

Costain, A. (1992). *Inviting women's rebellion: A political process interpretation of the women's movement.* Baltimore: Johns Hopkins University Press.

Curran, J. (1991). Rethinking the media as a public sphere. In P. Dahlgren & C. Sparks (Eds.), *Communication and citizenship.* London: Routledge.

Cushman, J. Jr. (1995, January 22). The Nation: Timber! A new idea is crashing. *New York Times,* Section 4, p. 5.

Dahlgren, P. (1985). The modes of reception: Towards a hermeneutics of TV news. In P. Drummond & R. Paterson (Eds.), *Television in transition* (pp. 235–249). London: British Film Institute.

Dahlgren, P., & Sparks, C. (Eds.). (1991). *Communication and citizenship: Journalism and the public sphere in the new media age.* London: Routledge.

de Certeau, M. (1988). *The practice of everyday life.* Berkeley and Los Angeles: University of California Press.

Debord, G. (1983). *Society of the spectacle.* Detroit: Black & Red.

Delgado, F. P. (1995). Chicano movement rhetoric: An ideographic interpretation. *Communication Quarterly, 43*(Fall), 446–454.

DeLuca, K. (1991). *Progress and the conceptualization of the environmental crisis.* Unpublished master's thesis, University of Massachusetts, Amherst.

Derrida, J. (1974/1976). *Of grammatology.* Baltimore: Johns Hopkins University Press.

Derrida, J. (1978a). Structure, sign, and play in the discourse of the human sciences. In *Writing and difference.* Chicago: University of Chicago Press.

Derrida, J. (1978b). *Writing and difference.* Chicago: University of Chicago Press.

Derrida, J. (1981a). *Dissemination.* Chicago: University of Chicago Press.

Derrida, J. (1981b). *Positions.* Chicago: University of Chicago Press.

Derrida, J. (1982). Signature event context. *Margins of philosophy* (pp. 307–330). Chicago: University of Chicago Press.

Derrida, J. (1987). *The post card: From Socrates to Freud and beyond.* Chicago: University of Chicago Press.

Derrida, J. (1988). *Limited inc.* Evanston, Il: Northwestern University Press.

Desilet, G. (1991). Heidegger and Derrida: The conflict between hermeneutics and deconstruction in the context of rhetorical and communication theory. *Quarterly Journal of Speech, 77,* 152–175.

Diamond, I., & Orenstein, G. F. (Eds.). (1990). *Reweaving the world: The emergence of ecofeminism.* San Francisco: Sierra Club Books.

Dobson, A. (Ed.). (1991). *The green reader.* San Francisco: Mercury House.

Dondis, D. A. (1993). *A primer of visual literacy.* Cambridge, MA: MIT Press.

Donovan, R. J., & Scherer, R. (1992). *Unsilent revolution: Television news and American public life.* Cambridge, UK: Cambridge University Press.

Douglas, M. (1975). *Implicit meanings: Essays in anthropology.* London: Routledge & Kegan Paul.

Dowie, M. (1995). *Losing ground: American environmentalism at the close of the twentieth century.* Cambridge, MA: MIT Press.

Eagleton, T. (1991). *Ideology, an introduction.* New York: Verso.

Easterbrook, G. (1989, July 24). Cleaning up our mess. *Newsweek,* pp. 27–42.

Egan, T. (1992, April 21). Chief's 1854 warning tied to 1971 ecological script. *New York Times,* pp. A1, A13.

Eley, G. (1992). Nations, publics, and political cultures: Placing Habermas in the nineteenth century. In C. Calhoun (Ed.), *Habermas and the public sphere* (pp. 289–339). Cambridge, MA: MIT Press.

Ellis, D. G. (1991). Post-structuralism and language: Non-sense. *Communication Monographs, 58,* 213–224.

Enzensberger, H. (1974). *The consciousness industry.* New York: Seabury Press.

Evernden, N. (1989). Nature in industrial society. In I. Angus & S. Jhally (Eds.), *Cultural politics in contemporary America* (pp. 151–164). New York: Routledge.

Evernden, N. (1992). *The social creation of nature*. Baltimore: Johns Hopkins University Press.

Fisher, W. (1987). *Human communication as narration*. Columbia: University of South Carolina Press.

Fiske, J. (1986). Television: Polysemy and popularity. *Critical Studies in Mass Communication, 3,* 391–408.

Fiske, J. (1989). *Understanding popular culture*. Cambridge, MA: Unwin Hyman.

Fiske, J. (1990). *Introduction to communication studies*. New York: Routledge.

Fiske, J. (1994). *Media matters*. Minneapolis: University of Minnesota Press.

Foreman, D. (1991a). *Confessions of an eco-warrior*. New York: Harmony Books.

Foreman, D. (1991b, Summer). The new conservation movement. *Wild Earth,* pp. 6–12.

Foreman, D., & Haywood, B. (Eds.). (1987). *Ecodefense: A field guide to monkeywrenching*. Tucson, AZ: Ned Ludd.

Foss, S. K., Foss, K. A., & Trapp, R. (1985). *Contemporary perspectives on rhetoric*. Prospect Heights, IL: Waveland Press.

Foucault, M. (1973). *The order of things*. New York: Vintage Books.

Foucault, M. (1978). *The history of sexuality* (Vol. 1). New York: Vintage Books.

Fraser, N. (1989). *Unruly practices: Power, discourse, and gender in contemporary social theory*. Minneapolis: University of Minnesota Press.

Fraser, N. (1992). Rethinking the public sphere: A contribution to the critique of actually existing democracy. In C. Calhoun (Ed.), *Habermas and the public sphere* (pp. 109–142). Cambridge, MA: MIT Press.

Frechet, G., & Worndl, B. (1993). The ecological movements in the light of social movements' development. *International Journal of Comparative Sociology, 34,* 56–74.

Gallup/CNN/USA Today Poll (1999, April 13–14). www.pollingreport.com/enviro.htm.

Gaonkar, D. P. (1990). Object and method in rhetorical criticism: From Wichelns to Leff and McGee. *Western Journal of Speech Communication, 54*(Summer), 290–316.

Gaonkar, D. P. (1993). The idea of rhetoric in the rhetoric of science. *Southern Communication Journal, 58*(4), 258–295.

Gaonkar, D. P. (1997). The idea of rhetoric in the rhetoric of science. In W. Keith & A. Gross (Eds.), *Rhetorical hermeneutics* (pp. 25–85). Albany: State University of New York Press.

Garnham, N. (1979). Contribution to a political economy of mass communication. *Media, Culture and Society, 1,* 126.

Garnham, N. (1995). Political economy and cultural studies: Reconciliation or divorce? *Critical Studies in Mass Communication 12:1,* 62–71.

Geras, N. (1987). Post-Marxism? *New Left Review, 163,* 40–82.

Gibbs, L. (1982). *Love Canal*. Albany: State University of New York Press.

Gibbs, L. (1993). Celebrating ten years of triumph. *Everyone's Backyard, 11,* 2.

Gitlin, T. (1980). *The whole world is watching*. Berkeley and Los Angeles: University of California Press.

Glacken, C. (1967). *Traces on the Rhodian shore*. Berkeley and Los Angeles: University of California Press.

Glaser, M. (1997, September 22). Censorious advertising. *Nation*, p. 7.

Glasser, T. L., & Ettema, J. L. (1989). Investigative journalism and the moral order. *Critical Studies in Mass Communication, 6*(1), 1–20.

Glasser, T. L., & Ettema, J. L. (1993). When the facts don't speak for themselves: A study of the uses of irony in daily journalism. *Critical Studies in Mass Communication, 10,* 322–338.

Goodell, J. (1999, January 21). Death in the redwoods. *rolling Stone*, pp. 60–69, 86.

Goodnight, G. T. (1982). The personal, technical, and public spheres of argument: A speculative inquiry into the art of public deliberation. *Journal of the American Forensic Association, 18,* 214–227.

Goodnight, G. T. (1987). Public discourse. *Critical Studies in Mass Communication, 4,* 428–432.

Goodnight, G. T. (1995). The firm, the park, and the university: Fear and trembling on the postmodern trail. *Quarterly Journal of Speech, 81,* 267–290.

Goody, J. (1977). *the domestication of the savage mind.* Cambridge, UK: Cambridge University Press.

Gottlieb, R. (1993). *Forcing the spring.* Washington, DC: Island Press.

Gouran, D. S. (Ed.). (1980). Social movements. *Central States Speech Journal, 31*(4), (Special Issue).

Gray, C. (1998, June 19). Louisian is nation's no. 2 polluter; No. 1 in amount of water waste. *The Times-Picayune*, p. A2.

The Greenpeace book. (1978). Vancouver, BC, Canada: Orca Sound.

Greenwire: The Environmental News Daily. (1998, June 3). Enviro groups: Sierra Club names Republican as new chief.

Gregg, R. B. (1971). The ego-function of the rhetoric of protest. *Philosophy and Rhetoric, 4*(2), 71–89.

Greider, W. (1988, December 15). The dirty politics of the environment. *Rolling Stone*, pp. 177–180.

Greider, W. (1992). *Who will tell the people?* New York: Simon & Schuster.

Griffin, L. M. (1952). The rhetoric of historical movements. *Quarterly Journal of Speech, 38,* 184–188.

Griffin, L. M. (1964). The rhetorical structure of the "New Left" movement: Part 1. *Quarterly Journal of Speech, 50,* 113–135.

Griffin, L. M. (1969). A dramatistic theory of the rhetoric of movements. In W. H. Rueckert (Ed.), *Critical response to Kenneth Burke, 1924–1966* (pp. 456–478). Minneapolis: University of Minnesota Press.

Griffin, L. M. (1980). On studying movements. *Central States Speech Journal, 31*(Winter), 225–232.

Griffin, S. (1978). *Woman and nature: The roaring inside her.* New York: Harper & Row.

Gronbeck, B. E. (Ed.). (1989). *Spheres of argument: Proceedings of the sixth SCA/AFA conference on argumentation.* Annandale, VA: Speech Communication Association.

Gronbeck, B. E. (1992). Negative narrative in 1988 presidential campaign ads. *Quarterly Journal of Speech, 78,* 333–346.

Gronbeck, B. E. (1995). Rhetoric, ethics, and telespectacles in the post-everything age. In R. H. Brown (Ed.), *Postmodern representations: Truth, power, and mimesis in the human sciences and public culture* (pp. 217–238). Urbana: University of Illinois Press.

Gronbeck, B. E., McKerrow, R. E., Ehninger, D., & Monroe, A. H. (1990). *Principles and types of speech communication.* Glenview, IL: Scott, Foresman.

Grosz, E. (1990). *Jacques Lacan: A feminist introduction.* London: Routledge.

Grosz, E. (1995). *Space, time, and perversion.* New York: Routledge.

Gup, T. (1990a, June 25). Artist with a 20-lb. saw. *Time,* p. 61.

Gup, T. (1990b, June 25). Owl vs. man. *Time,* pp. 56–65.

Habermas, J. (1974). The public sphere: An encyclopedic article (1964). *New German Critique, 1,* 49–55.

Habermas, J. (1987). *The philosophical discourse of modernity.* Cambridge, MA: MIT Press.

Habermas, J. (1989). *The structural transformation of the public sphere.* Cambridge, MA: MIT Press.

Habermas, J. (1992). Further reflections on the public sphere. In C. Calhoun (Ed.), *Habermas and the public sphere* (pp. 421–461). Cambridge, MA: MIT Press.

Haiman, F. S. (1967). The rhetoric of the streets. *Quarterly Journal of Speech, 53,* 99–114.

Hall, S. (1973). The determination of news photographs. In S. Cohen & J. Young (Eds.), *The manufacture of news.* London: Constable.

Hall, S. (1980a). Encoding/decoding. In S. Hall, D. Hobson, A. Lowe, & P. Willis (Eds.), *Culture, media, language* (pp. 128–138). London: Hutchinson.

Hall, S. (1980b). Race, articulation, and societies structured in dominance. In *Sociological theories: Race and colonialism* (pp. 305–345). Paris: UNESCO.

Hall, S. (1986). On postmodernism and articulation: An interview with Stuart Hall. *Journal of Communication Inquiry, 10,* 45–60.

Hall, S. (1994). Reflections upon the encoding/decoding model: An interview with Stuart Hall. In J. Cruz & J. Lewis (Eds.), *Viewing, reading, listening: Audiences and cultural reception* (pp. 253–274). Boulder, CO: Westview Press.

Hall, S., Hobson, D., Lowe, A., & Willis, P. (Eds.). (1980). *Culture, media, language.* London: Hutchinson.

Handler, J. F. (1992). Postmodernism, protest, and the new social movements. *Law & Society Review, 26,* 697–731.

Haraway, D. (1989). *Primate visions: Gender, race, and nature in the world of modern science.* New York: Routledge, Chapman, & Hall.

Haraway, D. (1991). *Simians, cyborgs, and women: The reinvention of nature.* New York: Routledge.

Hariman, R. (1991). Critical rhetoric and postmodern theory. *Quarterly Journal of Speech, 77,* 67–70.

Hariman, R. (1997). Afterword: Relocating the art of public address. In T. Benson (Ed.), *Rhetoric and political culture in nineteenth-century America* (pp. 163–183). East Lansing: Michigan State University Press.

Hart, R. P. (1994). *Seducing America: How television charms the modern voter.* New York: Oxford University Press.

Harvey, D. (1989). *The condition of postmodernity.* Cambridge, MA: Basil Blackwell.

Harwood, M. (1988, October 2). Daredevils for the environment. *New York Times*, pp. 72–76.

Hauser, G. A. (1987). Features of the public sphere. *Critical Studies in Mass Communication, 4*, 437–441.

Havelock, E. A. (1963). *The muse learns to write: Reflections on orality and literacy from antiquity to the present.* New Haven, CT: Yale University Press.

Hayles, N. K. (1990). *Chaos bound.* Ithaca, NY: Cornell University Press.

Heidegger, M. (1977). *The question concerning technology and other essays.* New York: Harper.

Helvarg, D. (1994a). *The war against the Greens: The Wise-use movement, the new right, and anti-environmental violence.* San Francisco: Sierra Club Books.

Helvarg, D. (1994b, November 28). The war on the Greens: Anti-enviros are getting uglier. *Nation*, pp. 646–651.

Helvarg, D. (1995a, January 30). "Property rights" movement: Legal assault on the environment. *Nation*, pp. 126–130.

Helvarg, D. (1995b, May 22). Property rights and militia: The anti-enviro connection. *Nation*, pp. 722–724.

Helvarg, D. (1995c, December 4). Defoliating our green laws: Congress plans an American clearcut. *Nation*, pp. 699–704.

Henry, D., & Jensen, R. J. (1991). Social movement criticism and the renaissance of public address. *Communication Studies, 42*(Spring), 83–93.

Herman, E. S., & Chomsky, N. (1988). *Manufacturing consent: The political economy of the mass media.* New York: Pantheon Books.

Hertsgaard, M. (1990, March–April). Covering the world: Ignoring the earth. *Greenpeace*, pp. 14–18.

Hikins, J. W. (1995). Nietzsche, Eristic, and the rhetoric of the possible: A commentary on the Whitson and Poulakos "Aesthetic view" of rhetoric. *Quarterly Journal of Speech, 81*, 353–377.

Hill, C. (1967). *Reformation to Industrial Revolution.* London: Weidenfeld & Nicolson.

Hill, C. (1972). *The world turned upside down: Radical ideas during the English Revolution.* New York: Viking Press.

Hill, C. (1985). *The collected essays of Christopher Hill.* Amherst: University of Massachusetts Press.

Hobsbawm, E. J. (1952). The machine breakers. *Past & Present, 1*(February), 57–70.

Hobsbawm, E. J. (1969). *Industry and empire.* Middlesex, UK: Penguin Books.

Hogan, J. M. (1996). *The nuclear freeze campaign, rhetoric and foreign policy in the telepolitical age.* East Lansing: Michigan State University Press.

Horkheimer, M. (1947). *Eclipse of reason.* New York: Oxford University Press.

Horkheimer, M. (1972). *Critical theory: Selected essays.* New York: Continuum.

Horkheimer, M., & Adorno, T. (1972). *Dialectic of enlightenment.* New York: Herder.

Hornblower, M. (1998, May 11). Five months at 180 ft.: An ecowarrior who calls herself Butterfly has set a tree-squatting record. *Time 151:18.*

Horton, T. (1991, September 5). The green giant. *Rolling Stone*, pp. 42–48, 108–112.

Hume, D. (1739/1962). *A treatise of human nature.* UK: Collins.

Hunter, R. (1971). *The storming of the mind.* Garden City, NY: Doubleday.

Hunter, R. (1979). *Warriors of the rainbow: A chronicle of the Greenpeace movement.* New York: Holt, Rinehart & Winston.

Hunter, R., & Weyler, R. (1978). *To save a whale: The voyages of Greenpeace.* San Francisco: Chronicle.

Hyde, M. J., & Smith, C. R. (1979). Hermeneutics and rhetoric: A seen but unobserved relationship. *Quarterly Journal of Speech, 65,* 347–363.

Hynes, T. J. Jr. (1989). *Can you buy cold fusion by the six pack? Or Bubba and Billy Bob discover Pons and Fleischmann.* In B. E. Gronbeck (Ed.), *Spheres of argument: Proceedings of the sixth SCA/AFA conference on argumentation* (pp. 42–46). Annandale, VA: Speech Communication Association.

Innis, H. A. (1950/1972). *Empire and communications.* London: Oxford University Press.

Innis, H. A. (1951/1964). *The bias of communication.* Toronto: University of Toronto Press.

Jacobson, L. (1998, May 29). Wilderness crusade. *Washington City Paper,* p. 48.

Jameson, F. (1991). *Postmodernism; or, The cultural logic of late capitalism.* Durham, NC: Duke University Press.

Jamieson, K. H. (1988). *Eloquence in an electronic age.* New York: Oxford University Press.

Jamieson, K. H. (1994, September 28). Political ads, the press, and lessons in psychology. *The Chronicle of Higher Education,* p. A56.

Jasinski, J. (1997). Instrumentalism, contextualism, and interpretation in rhetorical criticism. In W. Keith & A. Gross (Eds.), *Rhetorical hermeneutics* (pp. 25–85). Albany: State University of New York Press.

Jhally, S. (1989). The political economy of culture. In I. Angus & S. Jhally (Eds.), *Cultural politics in contemporary America* (pp. 65–81). New York: Routledge.

Jhally, S., & Lewis, J. (1992). *Enlightened racism, "The Cosby Show," audiences, and the myth of the American dream.* Boulder, CO: Westview Press.

Johnstone, C. J. (1974). Thoreau and civil disobedience: A rhetorical paradox. *Quarterly Journal of Speech, 60,* 313–322.

Kane, J. (1987, February). Mother Nature's army. *Esquire,* pp. 98–106.

Kaplan, T. (1997). *Crazy for democracy: Women in grassroots movements.* New York: Routledge.

Katz, E., & Dayan, D. (1986). Contents, conquests, and coronations: Media events and their heroes. In C. F. Graumann & S. Moscovici (Eds.), *Changing conceptions of leadership* (pp. 135–144). New York: Springer.

Kentuckians for the Commonwealth. *Fighting for justice: KFTC* [Film]. Whitesburg, KY: Appalshop Films.

Kenworthy, T. (1995, April 7). Wilderness Society president sold timber cut on his Montana ranch. *Washington Post,* p. A28.

Kerr, M. L., & Lee, C. (1993). From conquistadors to coalitions. *Southern Exposure, 21*(4), 8–19.

Kheel, M. (1990). Ecofeminism and deep ecology. In I. Diamond & G. F. Orenstein (Eds.), *Reweaving the world: The emergence of ecofeminism* (pp. 128–137). San Francisco: Sierra Club Books.

King, A. (1992). What is postmodern rhetoric? In A. King (Ed.), *Postmodern political communication* (pp. 1–12). Westport, CT: Praeger.

Kline, S. (1989). Limits to the imagination: Marketing and children's culture. In I. Angus & S. Jhally (Eds.), *Cultural politics in contemporary America* (pp. 299–316). New York: Routledge.

Laclau, E. (1990). *New reflections on the revolution of our time.* London: Verso.

Laclau, E. (1993a). Politics and the limits of modernity. In T. Docherty (Ed.), *Postmodernism: A reader* (pp. 329–343). New York: Columbia University Press.

Laclau, E. (1993b). Power and representation. In M. Poster (Ed.), *Politics, theory, and contemporary culture* (pp. 277–296). New York: Columbia University Press.

Laclau, E., & Mouffe, C. (1985). *Hegemony and socialist strategy: Towards a radical democratic politics.* London: Verso.

Lake, R. A. (1983). Enacting red power: The consummatory function in Native American protest rhetoric. *Quarterly Journal of Speech, 69,* 127–142.

Lancaster, J. (1991, March 20). The green guerrilla: "Redneck" eco-acivist Dave Foreman, throwing a monkey wrench into the system. *Washington Post,* p. B1.

Landes, J. (1988). *Women and the public sphere in the age of the French Revolution.* Ithaca, NY: Cornell University Press.

Lange, J. I. (1990). Refusal to compromise: The case of Earth First! *Western Journal of Speech Communication, 54*(Fall), 473–494.

Langsdorf, L. (1993). Words of others and sightings/citings/sitings of self. In I. Angus & L. Langsdorf (Eds.), *The critical turn: Rhetoric and philosophy in postmodern discourse* (pp. 20–50). Carbondale: Southern Illinois University Press.

Lanier, L. H. (1972). A critique of the philosophy of progress: The South and the agrarian tradition. In *I'll take my stand.* Baton Rouge: Louisiana State University Press.

Larson, C. U. (1989). *Persuasion.* Belmont, CA: Wadsworth.

Lasch, C. (1991). *The true and only heaven.* New York: Norton.

Leff, M. C. (1992). Things made by words: Reflections on textual criticism. *Quarterly Journal of Speech, 78,* 223–231.

Leff, M. (1997). Lincoln among the nineteenth-century orators. In T. Benson (Ed.), *Rhetoric and political culture in nineteenth-century America* (pp. 131–156). East Lansing: Michigan State University Press.

Leff, M. C., & Sachs, A. (1990). Words the most like things: Iconicity and the rhetorical text. *Western Journal of Speech Communication, 54*(Summer), 252–273.

Leiss, W. (1974). *The domination of nature.* Boston: Beacon Press.

Leiss, W. (1989). The myth of the information society. In I. Angus & S. Jhally (Eds.), *Cultural polities in contemporary America* (pp. 96–107). New York: Routledge.

Leonard, J. (1993, May 17). Machine dreams. *Nation,* pp. 667–672.

Leopold, A. (1949/1968). *A Sand County almanac.* Oxford, UK: Oxford University Press.

Levine, M. (1982). Introduction. In L. Gibbs (Ed.), *Love Canal* (pp. xiii–xvii). Albany: State University of New York Press.

Lewis, J. (1981/82). The story of a riot. *Screen Education, 40,* 15–33.

Lewis, J. (1985). Decoding television news. In P. Drummond & R. Paterson (Eds.), *Television in transition* (pp. 205–234). London: British Film Institute.

Lewis, J. (1991). *The ideological octopus: An exploration of television and its audience.* New York: Routledge.

Lewis, J. (1994). The meaning of things: Audiences, ambiguity, and power. In J. Cruz & J. Lewis (Eds.), *Viewing, reading, listening: Audiences and cultural reception* (pp. 19–32). Boulder, CO: Westview Press.

Li, H. (1993). A cross-cultural critique of ecofeminism. In G. Gaard (Ed.), *Ecofeminism.* Philadelphia: Temple University Press.

Linden, E. (1989, January 2). The death of birth. *Time,* pp. 32–35.

Lingis, A. (1994). *Foreign bodies.* New York: Routledge.

Lucaites, J. L. (1997). Visualizing "The People": Individualism vs. collectivism in *Let us now praise famous men. Quarterly Journal of Speech, 83,* 269–288.

Lucas, S. E. (1988). The renaissance of American public address: Text and context in rhetorical criticism. *Quarterly Journal of Speech, 74,* 241–260.

Lucas, S. E. (1995). *The art of public speaking.* New York: McGraw-Hill.

Luoma, J. R. (1991, November–December). Right in your own backyard. *Audubon,* pp. 88–95.

Lyotard, J. (1979). *The postmodern condition: A report on knowledge.* Minneapolis: University of Minnesota Press.

Madison, I. (1993, Winter). To forge a movement. *Southern Exposure,* pp. 30–33.

Mander, J. (1991). *In the absence of the sacred: The failure of technology and the survival of the Indian nations.* San Francisco: Sierra Club Books.

Manes, C. (1990). *Green rage, radical environmentalism, and the unmaking of civilization.* Boston: Little, Brown.

Manoff, R., & Schudson, M. (1986). *Reading the news.* New York: Pantheon Books.

Marcuse, H. (1964). *One dimensional man.* Boston: Beacon Press.

Marsh, G. P. (1864). *Man and nature; or, Physical geography as modified by human action.* New York: Scribner.

McDonald, S. (1993, September). The media and misinformation: How the press has fueled an anti-environmental backlash. *Friends of the Earth,* pp. 6–7.

McEdwards, M. G. (1968). Agitative rhetoric: Its nature and effect. *Western Speech, 32,* 36–43.

McGee, M. C. (1975). In search of "the people": A rhetorical alternative. *Quarterly Journal of Speech, 61,* 235–249.

McGee, M. C. (1980a). The "ideograph": A link between rhetoric and ideology. *Quarterly Journal of Speech, 66,* 1–16.

McGee, M. C. (1980b). The origins of liberty: A feminization of power. *Communication Monographs, 47,* 23–45.

McGee, M. C. (1980c). "Social movement": Phenomenon or meaning? *Central States Speech Journal, 31,* 233–234.

McGee, M. C. (1983). Social movement as meaning. *Central States Speech Journal, 34,* 74–77.

McGee, M. C. (1990). Text, context, and the fragmentation of contemporary culture. *Western Journal of Speech Communication, 54,* 274–289.

McKerrow, R. E. (1989). Critical rhetoric: Theory and praxis. *Communication Monographs, 56,* 91–111.

McKerrow, R. E. (1991a). Critical rhetoric in a postmodern world. *Quarterly Journal of Speech, 77,* 75–78.

McKerrow, R. E. (1991b). Critical rhetoric and propaganda studies. *Communication Yearbook, 14,* 249–255.

McKerrow, R. E. (1993). Critical rhetoric and the possibility of the subject. In I. Angus & L. Langsdorf (Eds.), *The critical turn: Rhetoric and philosophy in postmodern discourse* (pp. 51–67). Carbondale: Southern Illinois University Press.

McKibben, B. (1989). *The end of nature.* New York: Random House.

McLaughlin, L. (1993). Feminism, the public sphere, media, and democracy. *Media, Culture and Society, 15,* 599–620.

McLuhan, M. (1962). *The Gutenberg galaxy: The making of typographic man.* Toronto: University of Toronto Press.

McLuhan, M. (1964). *Understanding media: The extension of man.* New York: Mc-Graw-Hill.

McLuhan, M. (1977, Autumn). The rise and fall of nature. *Journal of Communication, 27,* 80–81.

McLuhan, T. C. (Ed.). (1971). *Touch the earth.* New York: Outerbridge & Dienstfrey.

McMahan, E. M. (1976). Nonverbal communication as a function of attribution in impression formation. *Communication Monographs, 43,* 287–294.

Medhurst, M. J., & Benson, T. (Eds.). (1991). *Rhetorical dimensions in media: A critical casebook* (2d ed.). Dubuque, IA: Kendall/Hunt.

Mehrabian, A. (1967). Orientation behaviors and nonverbal attitude communication. *Journal of Communication, 17,* 324–332.

Melucci, A. (1978, February). Ten hypotheses for the analysis of new movements. *Quaderni Piacentini, 65,* 3–19.

Melucci, A. (1980). The new social movements: A theoretical approach. *Social Science Information, 19,* 199–226.

Melucci, A. (1984, April–May). An end to social movements? Introductory paper to the "Sessions on New Movements and Change in Organizational Forms." *Social Science Information, 23,* 819–835.

Melucci, A. (1985, Winter). The symbolic challenge of contemporary movements. *Social Research, 52,* 789–816.

Merchant, C. (1980). *The death of nature: Women, ecology, and the scientific revolution.* San Francisco: Harper & Row.

Merchant, C. (1990). Ecofeminism and feminist theory. In I. Diamond & G. F. Orenstein (Eds.), *Reweaving the world: The emergence of ecofeminism* (pp. 100–105). San Francisco: Sierra Club Books.

Merchant, C. (1991). Women and nature. In A. Dobson (Ed.), *The green reader* (pp. 258–261). San Francisco: Mercury House.

Merchant, C. (1992). *Radical ecology: The search for a livable world.* New York: Routledge.

Merleau-Ponty, M. (1962). *Phenomenology of perception.* New York: Humanities Press.

Meyrowitz, J. (1985). *No sense of place: The impact of electronic media on social behavior.* New York: Oxford University Press.

Miller, M. C. (1996, June 3). Free the media. *Nation,* pp. 9–15.

Miller, M. C. (1998, June 8). TV: The nature of the beast. *Nation,* pp. 11–13.

Miller, M. (n.d.). *Spectacle.*

Mitchell, S. (1996). *The official guide to American attitudes.* Ithaca, NY: New Strategist Publications, Inc.

Montague, P. (1993). After ten years: Reason for hope. In Movement on the Move, *Ten years of triumph* (pp. 14–15). United States: Citizen's Clearinghouse for Hazardous Wastes.

Morin, R., & Balz, D. (1996, January 28,). In America, loss of confidence seeps into all institutions: Suspicion of strangers breeds widespread cynicism. *Washington Post,* pp. A1, A6–7.

Morley, D. (1980a). *The "nationwide" audience: Structure and decoding.* London: British Film Institute.

Morley, D. (1980b). Texts, readers, subjects. In S. Hall, D. Hobson, A. Low, & P. Willis (Eds.), *Culture, media, language* (pp. 163–176). London: Hutchinson.

Morley, D. (1981). The nationwide audience: A critical postscript. *Screen Education, 39,* pp. 3–14.

Mouffe, C. (1992). Feminism, citizenship, and radical democratic politics. In J. Butler & J. Scott (Eds.), *Feminists theorize the political* (pp. 369–384). New York: Routledge.

Movement on the Move. (1993). *Ten years of triumph.* United States: Citizen's Clearinghouse for Hazardous Wastes.

Moyers, B. D. (1989). *The public mind: Image and reality in America.* Alexandria, VA: PBS Video.

Muir, J. (1967/1968). *Gentle wilderness: The Sierra Nevada.* New York: Ballantine Books.

Mumford, L. (1934). *Technics and civilization.* New York: Harcourt, Brace.

Naess, A. (1973). The shallow and the deep, long-range ecology movement: A summary. *Inquiry, 16,* 101–137.

Nakayama, T. K., & Krizek, R. L. (1995). Whiteness: A strategic rhetoric. *Quarterly Journal of Speech, 81,* 291–309.

Nash, R. (1967/1973). *Wilderness and the American mind* (Rev. ed.). Binghamton, NY: Vail-Ballou Press.

National Public Radio. (1993, July 25). Earth First! Group locates to Idaho to stop loggers. *All Things Considered.*

National Safety Council. (1995). *Accident facts.* Itasca, IL: National Safety Council.

NBC News.(1975, July 1). New York: NBC.

NBC News.(1990, July 5,). Assignment earth: Earth First! New York: NBC.

NBC News.(1990, July 20). New York: NBC.

Nelson, J., Megill, A., & McCloskey, D. (Eds.). (1987). *The rhetoric of the human sciences: Language and argument in scholarship and public affairs.* Madison: University of Wisconsin Press.

Nisbet, R. (1980). *History of the idea of progress.* New York: Basic Books.

Noble, D. F. (1977). *America by design: Science, technology, and the rise of corporate capitalism.* New York: Knopf.

Noble, D. F. (1984). *Forces of production: A social history of industrial automation*. New York: Knopf.

Nocera, J. (1995, June). The profit motive: The gall of Goldsmith. *GQ*, 73–76.

Norris, C. (1992). *Uncritical theory, postmodernism, intellectuals, and the Gulf War*. London: Lawrence & Wishart.

Oelschlaeger, M. (1991). *The idea of wilderness*. New Haven, CT: Yale University Press.

Offe, C. (1985, Winter). New social movements: Challenging the boundaries of institutional politics. *Social Research, 52*, 817–868.

Olson, K. M., & Goodnight, G. T. (1989). Epochal rhetoric in 19th-century America: On the discursive instantiation of the technical sphere. In B. E. Gronbeck (Ed.), *Spheres of argument: Proceedings of the sixth SCA/APA conference on argumentation*. Annandale, VA: Speech Communication Association.

Olson, K. M., & Goodnight, G. T. (1994). Entanglements of consumption, cruelty, privacy, and fashion: The social controversy of fur. *Quarterly Journal of Speech, 80*, 249–276.

Olson, L. C. (1991). *Emblems of American community in the revolutionary era: A study in rhetorical iconology*. Washington, DC: Smithsonian Institution Press.

Ong, W. J. (1982). *Orality and literacy: The technologizing of the word*. London: Metheun.

Ono, K. A., & Sloop, J. M. (1992, March). Commitment to *telos*—A sustained critical rhetoric. *Communication Monographs, 59*, 48–60.

Ortner, S. (1974). Is female to male as nature is to culture? In M. Rosaldo & L. Lamphere (Eds.), *Woman, culture, and society* (pp. 67–87). Palo Alto, CA: Stanford University Press.

Orwin, C. (1996, Spring). Distant compassion: CNN and Borrioboola-Gha. *National Interest*, pp. 42–49.

O'Shaughnessy, K. (Producer/director). (1993). *My name is Allegany* [Film]. Buffalo, NY: Hallways Contemporary Arts Center.

Ostertag, B. (1991, March–April). Greenpeace takes over the world. *Mother Jones*, pp. 32–34, 37, 82–87.

Paepke, C. O. (1993). *The evolution of progress, the end of economic growth and the beginning of human transformation*. New York: Random House.

Papazian, E. (Ed.). (1995). *TV dimensions '95*. New York: Media Dynamics.

Parenti, M. (1993). *Inventing reality*. New York: St. Martin's Press.

Parfit, M. (1990, April). Earth First-ers! wield a mean monkey wrench. *Smithsonian*, pp. 184–204.

Pateman, C. (1988). The fraternal social contract. In J. Keane (Ed.), *Civil society and the state: New European perspectives* (pp. 101–127). London: Verso.

Patton, C. (1990). *Inventing AIDS*. New York: Routledge.

PBS. (1990, July 20). *Focus—Logjam*: MacNeil–Lehrer Newshour.

Pell, E. (1990, April 16). Mad money: Is the environmental movement selling its soul for grants from polluters? *Valley Advocate*, pp. 13–14.

Peters, J. D. (1989). John Locke, the individual, and the origin of communication. *Quarterly Journal of Speech, 75*, 387–399.

Peters, J. D. (1993). Distrust of representation: Habermas on the public sphere. *Media, Culture and Society, 15*, 541–571.

Peters, J. D. (1994). The gaps of which communication is made. *Critical Studies in Mass Communication, 11*, 117–140.

Peters, J. D. (1996). Sharing thoughts or recognizing otherness? *Critical Studies in Mass Communication 13:4* pp. 373–380.

Peters, J. D. (n.d.). *The object of communication theory.* Unpublished manuscript at the University of Iowa.

Peters, J. D., & Cmiel, K. (1991). Media ethics and the public sphere. *Communications, 12*(3), 197–215.

Peterson, T. R. (1988). The rhetorical construction of institutional authority in a senate subcommittee hearing on wilderness legislation. *Western Journal of Speech Communication, 52,* 259–276.

Piller, C. (1992, July–August). Nimbymania. *Utne Reader,* pp. 114–115.

Porritt, J. (1984). *Seeing green: The politics of ecology explained.* Oxford, UK: Basil Blackwell.

Posner, M. I., Nissen, M. J., & Klein, R. M. (1976). Visual dominance: An information-processing account of its origins and significance. *Psychological Review, 83,* 157–171.

Poster, M. (1990). *The mode of information: Poststructuralism and social context.* Cornwall, UK: Polity Press.

Postman, N. (1985). *Amusing ourselves to death: Public discourse in the age of show business.* New York: Viking Press.

Powell, K. (1992). *Lifestyle as a dimension of social movement study: A case study of the vegetarian movement in the United States.* Unpublished PhD dissertation, University of Georgia, Athens.

Powell, K. (1995). The Association of Southern Women for the Prevention of Lynching: Strategies of a movement in the comic frame. *Communication Quarterly, 43,* 86–99.

Principles, The. (1993). *Southern Exposure, 21*(4), 19.

Public Agenda Online. (1999, May 25). Environment: People's chief concerns. www.publicagenda.org/issues.

Quinby, L. (1990). Ecofeminism and the politics of resistance. In I. Diamond & G. F. Orenstein (Eds.), *Reweaving the world: The emergence of ecofeminism* (pp. 122–127). San Francisco: Sierra Club Books.

Rifkin, J. (1980). *Entropy: A new world view.* New York: Viking Press.

Robbins, J. (1988, March 27). The environmental guerrillas. *Boston Globe,*

Robbins, J. (1991, April 19–21). Are cowboys killing the West? *USA Weekend,* pp. 4–6.

Romano, C. (1986). What? The grisly truth about bare facts. In R. Manoff & M. Schudson (Eds.), *Reading the news* (pp. 38–78). New York: Pantheon Books.

Rose, M. (1993). *Authors and owners.* Cambridge, MA: Harvard University Press.

Rossney, R. (1994, Spring). The new old Luddites: What's so funny about staying alive? *Whole Earth, 82,* 3–12.

Rowell, A. (1996). *Green backlash: Global subversion of the environmental movement.* New York: Routledge.

Ryan, M. (1992). Gender and public access: Women's politics in nineteenth-century America. In C. Calhoun (Ed.), *Habermas and the public sphere* (pp. 259–288). Cambridge, MA: MIT Press.

Safire, W. (1996, February 12). Chunnel vision. *New York Times*, pp. A15.

Sale, K. (1986, November). The forest for the trees. *Mother Jones*, pp. 25–58.

Sale, K. (1993). *The green revolution: The American environmental movement, 1962–1992*. New York: Hill & Wang.

Sale, K. (1995). *Rebels against the future: The Luddites and their war on the Industrial Revolution*. Reading, MA: Addison-Wesley.

Sancton, T. A. (1989, January 2). What on earth are we doing? *Time*, pp. 26–30.

Sawhill, J. C. (1990, June 9). What good are pupfish and periwinkles. *New York Times*, p. 23.

Scarce, R. (1990). *Eco-warriors: Understanding the radical environmental movement*. Chicago: Noble Press.

Schiappa, E. (1989). Spheres of argument as *topoi* for the critical study of power/knowledge. In B. E. Gronbeck (Ed.), *Spheres of argument: Proceedings of the sixth SCA/AFA conference on argumentation* (pp. 47–56). Annandale, VA: Speech Communication Association.

Schiller, H. (1989). The privatization of culture. In I. Angus & S. Jhally (Eds.), *Cultural politics in contemporary America* (pp. 317–332). New York: Routledge.

Schlechtweg, H. (1990, April 25). *Engaging dialogues: Rhetoric, philosophy, and post-Marxist politics*. Paper delivered at a philosophy symposium held at the Rochester Institute of Technology, Rochester, NY.

Schwab, J. (1994). *Deeper shades of green: The rise of blue-collar and minority environmentalism in America*. San Francisco: Sierra Club Books.

Scott, R., & Smith, D. (1969). The rhetoric of confrontation. *Quarterly Journal of Speech, 58*, 1–8.

Searle, J. R. (1977). Reiterating the differences: A reply to Derrida. *Glyph: John Hopkins Textual Studies*, Vol. 1, 198–208. Baltimore: Johns Hopkins University Press.

Seideman, D. (1990, June 25,). Terrorist in a white collar. *Time, p. 60.*

Seiter, E. (1987). Semiotics and television. In R. Allen (Ed.), *Channels of discourse* (pp. 17–41). Chapel Hill: University of North Carolina Press.

Setterberg, F. (1987, May–June). The wild bunch: Earth First! shakes up the environmental movement. *Utne Reader*, pp. 68–76.

Shaiko, R. G. (1993). Greenpeace U.S.A.: Something old, new, borrowed. *American Association of Political and Social Science, 528*, 88–100.

Sharkey, J. (1993, December). When pictures drive foreign policy: Somalia raises serious questions about media influence. *American Journalism Review, 15:10*, 14–19.

Sheedy, K. (n. d.). *Developing a rhetorical theory of social movements: A review and revision of the establishment-conflict model*. Unpublished manuscript, University of Georgia, Athens.

Shepard, K., & McKinley, D. (Eds.). (1969). *The subversive science: Essays toward an ecology of man*. Boston: Houghton Mifflin.

Shiva, B. (1988). *Staying alive*. London: Zed Books.

Short, B. (1991). Earth First! and the rhetoric of moral confrontation. *Communication Studies, 42*, 172–188.

Simons, H. W. (1970, February). Requirements, problems, and strategies: A theory of persuasion for social movements. *Quarterly Journal of Speech, 56*, 1–11.

Simons, H. W. (1972). Persuasion in social conflicts: A critique of prevailing conceptions and a framework for future research. *Speech Monographs, 39,* 227–247.

Simons, H. W. (1976, December). Changing notions about social movements. *Quarterly Journal of Speech, 62,* 425–430.

Simons, H. W. (1980, Winter). On terms, definitions, and theoretical distinctiveness: Comments on papers by McGee and Zarefsky. *Central States Speech Journal, 31,* 306–315.

Simons, H. W. (Ed.). (1989). *Rhetoric in the human sciences.* Newbury, CA: Sage.

Simons, H. W. (1991). On the rhetoric of social movements, historical movements, and "top-down" movements: A commentary. *Communication Studies, 42*(Spring), 94–101.

Simons, H. W., Mechling, E. W., & Schreier, H. N. (1984). The functions of human communication in mobilizing for action from the bottom up: The rhetoric of social movements. In C. C. Arnold & J. W. Bowers (Eds.), *Handbook of rhetorical and communication theory* (pp. 792–867). Boston: Allyn & Bacon.

60 Minutes. (1993, June 6). *Clear air, clean water, dirty fight.* New York: CBS.

Smith, M. (1997, Winter). Against the enclosure of the ethical commons: Radical environmentalism as an "ethics of place." *Environmental Ethics, 18,* 339–353.

Smith, N. (1984). *Uneven development, nature, capital, and the production of space.* Cambridge, MA: Basil Blackwell.

Smythe, D. (1981). *Dependency road: Communications, capitalism, consciousness, and Canada.* Norwood, NJ: Ablex.

Snyder, G. (1990). *The practice of the wild.* San Francisco: North Point Press.

Soja, E. (1989). *Postmodern geographies: The reassertion of space in critical social theory.* London: Verso.

Sontag, S. (1977). *On photography.* New York: Farrar, Straus, & Giroux.

Soule, M. & Lease, G. (1995). *Reinventing nature? Responses to postmodern deconstruction.* Washington, DC: Island Press.

Spencer, L., Bollwerk, J., & Morais, R. C. (1991, November 11). The not so peaceful world of Greenpeace. *Forbes,* pp. 174–180.

Spivak, G. C. (1976). Translator's preface. In J. Derrida (Ed.), *Of grammatology* (pp. ix–xc). Baltimore: Johns Hopkins University Press.

Stabile, C. (1994). "A garden inclosed is my sister": Ecofeminism and eco-valences. *Cultural Studies, 8*(1), 56–73.

Stent, G. S. (1978). *Paradoxes of progress.* San Francisco: Freeman.

Stewart, C. J. (1983). The functional perspective on the study of social movements. *Central States Speech Journal, 34*(Spring), 77–79.

Stewart, C. J. (1991). The internal rhetoric of the Knights of Labor. *Communication Studies, 42,* 67–82.

Stewart, C. J., Smith, C., & Denton, R. J. (1989). *Persuasion and social movements.* Prospect Heights, IL: Waveland.

Szasz, A. (1995). *Ecopopulism: Toxic waste and the movement for environmental justice.* Minneapolis: University of Minnesota Press.

Talbot, S. (1990, November–December). Earth First! What next? *Mother Jones,* pp. 15, 46–49.

Tatum, T.P.D. (1988). *Greenpeace's greatest hits* [Video]. Washington, DC: Tatum Video and Alpha Studios.

Taylor, B. (1991, November–December). The religion and politics of Earth First! *Ecologist, 21,* 258–266.

Thompson, E. P. (1964). *The making of the English working class.* New York: Pantheon Books.

Thoreau, H. D. (1948/1966). *Walden & On the duty of civil disobedience.* New York: Holt, Rinehart, & Winston.

Tiemens, R. K. (1989). The visual context of argument: An analysis of the September 25, 1988, presidential debate. In B. Gronbeck (Ed.), *Spheres of argument: Proceedings of the sixth SCA/AFA conference on argumentation* (pp. 140–146). Annandale, VA: Speech Communication Association.

Toulmin, S. (1990). *Cosmopolis: The hidden agenda of modernity.* New York: Free Press.

Touraine, A. (1985). An introduction to the study of social movements. *Social Research, 52*(Winter), 749–787.

Tubbs, S. L., & Moss, S. (1983). *Human communication.* New York: Random House.

Tuchman, G. (1978). *Making news.* Glencoe, NY: Free Press.

Ulmer, G. (1981). The post-age (Vol. 2). New York: *Diacritics.*

Ulmer, G. (1989). *Teletheory: Grammatology in the age of video.* New York: Routledge.

Van Gelder, L. (1992, January–February). Saving the homeplace: How Kentucky's most powerful environmental group gets what it wants. *Audubon,* pp. 62–67.

Vanderpool, T. (1989, September). Monkey-wrenching for Planet Earth. *Progressive,* pp. 53, 15.

Wald, M. L. (1990, April 22,). Guarding environment: A world of challenges. *New York Times,* pp. 1, 24–25.

Wallace, A. (1993). *Eco-heroes.* San Francisco: Mercury House.

Wallinger, M. J. (1989). Regulatory rhetoric: Argument in the nexus of public and technical spheres. In B. E. Gronbeck (Ed.), *Spheres of argument: Proceedings of the sixth SCA/AFA conference on argumentation* (pp. 71–80). Annandale, VA: Speech Communication Association.

Warnick, B. (1989). Judgement, probability, and Aristotle's *Rhetoric. Quarterly Journal of Speech, 73,* 299–311.

Warnick, B. (1992, May). Leff in context: What is the critic's role? *Quarterly Journal of Speech, 78,* 232–237.

Warren, K. (1987). Feminism and ecology: Making connections. *Environmental Ethics, 9*(Spring), 3–20.

Watson, B. (1994). For a while, the Luddites had a smashing success. *Smithsonian,* pp. 140–154.

Webster, D. (1992, July–August). Sweet home Arkansas. *Utne Reader,* pp. 112–116.

White, L. J. (1968). *Machino ex deo: Essays in the dynamism of Western culture.* Cambridge, MA: MIT Press.

White, S. K. (1991). *Political theory and postmodernism.* Cambridge University Press.

Wilbon, M. (1995, May 7). Profit over purity. *Washington Post,* pp. D1, D4.

Wilkinson, C. A. (1976). A rhetorical definition of movements. *Central States Speech Journal, 27*(Summer), 88–94.

Williams, M. R. (1994). A reconceptualization of protest rhetoric: Implications for the study of movements. *Women's Studies in Communication, 17*, 20–44.

Williams, R. (1977). *Marxism and literature.* New York: Oxford University Press.

Williams, R. (1980). *Problems in materialism and culture.* Great Britain: Redwood Burn.

Windt, T. O. (1972). The diatribe: Last resort for protest. *Quarterly Journal of Speech, 58*, 1–14.

Winner, L. (1986). *The whale and the reactor.* Chicago: University of Chicago Press.

Zarefsky, D. (1977). President Johnson's War on Poverty: The rhetoric of three "establishment" movements. *Communication Monographs, 40*, 159–165.

Zarefsky, D. (1980). A skeptical view of movement studies. *Central States Speech Journal, 31*, 245–254.

Zuercher, M. (1991). *Making history: The first ten years of KFTC.* Whitesbury, KY: Kentuckians for the Commonwealth.

Index

ABOUT THE AUTHOR

Kevin Michael DeLuca, PhD, has taught at the University of Virginia and the Pennsylvania State University and is currently an assistant professor of Speech Communication at the University of Georgia. His major area of interest is how industrial cultures relate to the natural world and construct visions of "nature." He has published articles on environmental politics, technology, the media, and postmodernism.